Cambridge Primary
Mathematics

Second Edition

Workbook 3

Catherine Casey
Steph King
Josh Lury

Series editors:
Paul Broadbent
Mike Askew

Boost

HODDER
EDUCATION
AN HACHETTE UK COMPANY

Acknowledgements

The Publishers would like to thank the following for permission to reproduce copyright material.

Photo credits
p. 7 *tl, cr,* **p. 13** *tl, cr,* **p. 19** *tl, cr,* **p. 23** *tl, cr,* **p. 29** *tl, cr,* **p. 33** *tl, cr,* **p. 41** *tl, cr,* **p. 46** *tl, cr,* **p. 52** *tl, cr,* **p. 56** *tl, cr,* **p. 62** *tl, cr,* **p. 66** *tl, cr,* **p. 71** *tl, cr,* **p. 78** *tl, cr,* **p. 82** *tl, cr,* **p. 88** *tl, cr,* **p. 92** *tl, cr,* **p. 96** *tl, cl* © Stocker Team/Adobe Stock Photo; **p. 53** *cl* © Sarawuth 123/Adobe Stock Photo; **p. 53** *bl* © Stock Photo-graf/Adobe Stock Photo; **p. 53** *cr* © Goir/Adobe Stock Photo; **p. 53** *br* © Voravuth/Adobe Stock Photo.

t = top, *b* = bottom, *l* = left, *r* = right, *c* = centre

Every effort has been made to trace all copyright holders, but if any have been inadvertently overlooked, the Publishers will be pleased to make the necessary arrangements at the first opportunity.

Hachette UK's policy is to use papers that are natural, renewable and recyclable products and made from wood grown in well-managed forests and other controlled sources. The logging and manufacturing processes are expected to conform to the environmental regulations of the country of origin.

Orders: please contact Hachette UK Distribution, Hely Hutchinson Centre, Milton Road, Didcot, Oxfordshire, OX11 7HH. Telephone: +44 (0)1235 827827. Email education@hachette.co.uk Lines are open from 9 a.m. to 5 p.m., Monday to Friday. You can also order through our website: www.hoddereducation.com

ISBN: 978 1 3983 0118 4

© Catherine Casey, Steph King and Josh Lury 2021

First published in 2017

This edition published in 2021 by

Hodder Education,

An Hachette UK Company

Carmelite House

50 Victoria Embankment

London EC4Y 0DZ

www.hoddereducation.com

Impression number   10 9 8 7 6
Year   2025 2024 2023

Cover illustration by Lisa Hunt, The Bright Agency

Illustrations by Jeanne du Plessis, James Hearne, Stéphan Theron, Natalie and Tamsin Hinrichsen

Typeset in FS Albert 15/17 by IO Publishing CC

Printed in Spain

A catalogue record for this title is available from the British Library.

MIX
Paper | Supporting
responsible forestry
FSC™ C104740

Contents

Term 1

Unit 1	Numbers to 1000	4
Unit 2	Addition and subtraction	8
Unit 3	Shapes and angles	14
Unit 4	Statistical methods and chance	20
Unit 5	Multiplication and division	24
Unit 6	Time and measurement	30

Term 2

Unit 7	Addition and subtraction	34
Unit 8	Patterns, place value and rounding	42
Unit 9	Multiplication and division	47
Unit 10	Time and measurement	53
Unit 11	Shapes and angles	57
Unit 12	Fractions	63

Term 3

Unit 13	Patterns, place value and rounding	67
Unit 14	Addition and subtraction	72
Unit 15	Time and measurement	79
Unit 16	Multiplication and division	83
Unit 17	Fractions	89
Unit 18	Statistical methods and chance	93

Remember: When you see this star ⭐, it is showing you that the activity develops your Thinking and Working Mathematically skills!

Can you remember?

What number does each diagram show? Write the value of the digit 4 each time.

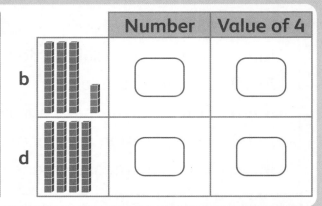

	Number	Value of 4
a		
c		

	Number	Value of 4
b		
d		

Numbers to 1000

1 Draw a diagram to show each number.

	a	b
113 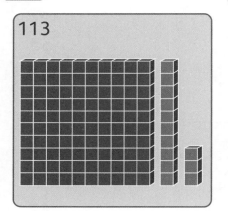	141	122

2 Draw lines to match each place value chart to its diagram.

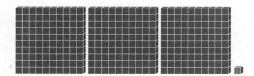

100s	10s	1s
3	1	1

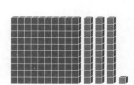

100s	10s	1s
1	3	1

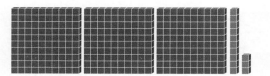

100s	10s	1s
3	1	3

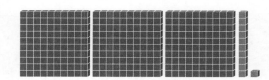

100s	10s	1s
3	0	1

3 Read each number. Then write the digits.

a One hundred and thirty-four =

b Three hundred and forty-one =

c Four hundred and thirty-one =

d Four hundred and thirteen =

e Four hundred and three =

f Three hundred and forty =

4 Sort these numbers into the table.

24 25 47 48 60 59 104 105 159 160 200 201 199

Odd numbers	Even numbers

Counting to 1000

1 Write the number at each arrow.

a
0 ... 1000

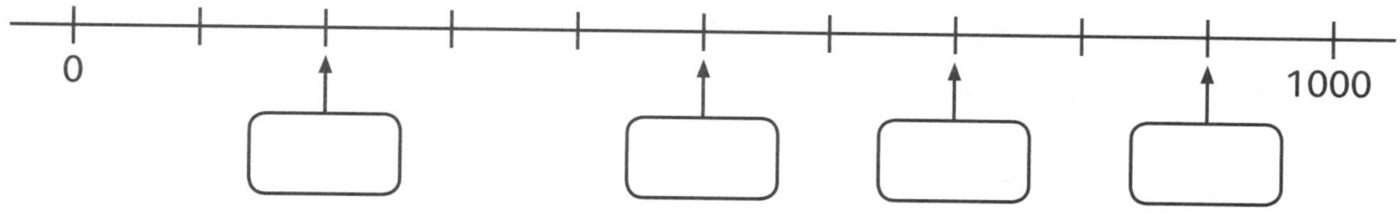

b
0 ... 1000

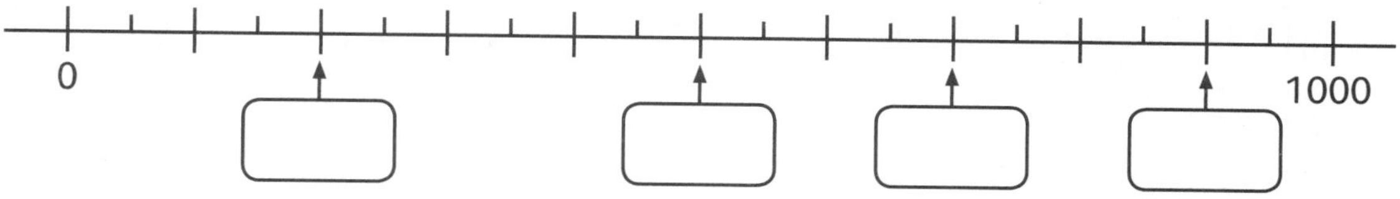

c
0 ... 100

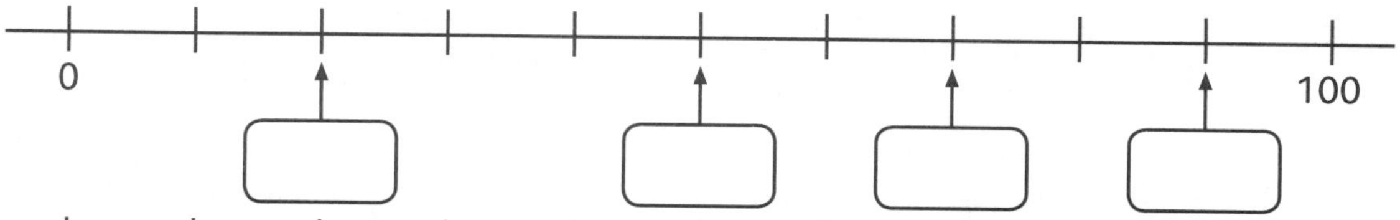

d
200 ... 300

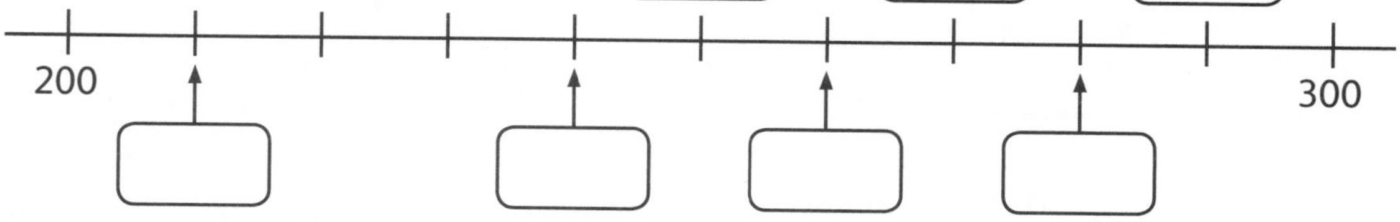

e
800 ... 900

5

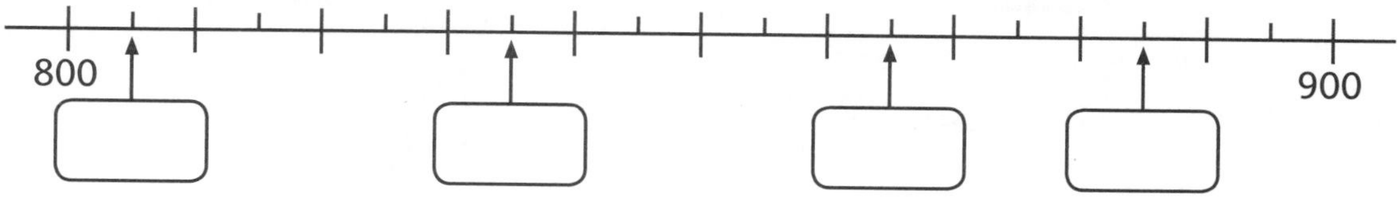

2 Complete each counting pattern.

a

5	15			45			75		

b

205	305	405							

c

950									860

d

955									55

3 Write the missing numbers.

a
+ 100   + 100   + 100

134 ⬚ ⬚ ⬚

b
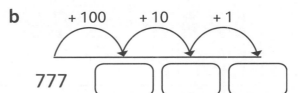
+ 100   + 10   + 1

777 ⬚ ⬚ ⬚

c
+ 100   + 100   + 100

431 ⬚ ⬚ ⬚

d
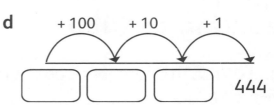
+ 100   + 10   + 1

⬚ ⬚ ⬚ 444

e
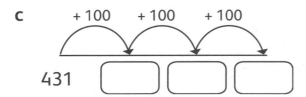
+ 100   + 100   + 100

⬚ ⬚ ⬚ 555

f
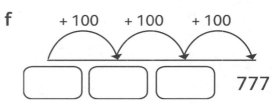
+ 100   + 100   + 100

⬚ ⬚ ⬚ 777

4 Fill in the missing numbers.

a [85] +100 → [ ]   b [95] +100 → [ ]

c [195] +100 → [ ]   d [ ] +100 → [395]

e [ ] +100 → [805]   f [ ] −100 → [905]

g [211] −100 → [ ]   h [210] −100 → [ ]

i [ ] −100 → [401]   j [401] −100 → [ ]

Unit 1 — Numbers to 1000

Self-check

 I can do this.

 I can do this, but I need to keep trying.

 I can't do this yet.

See how much you know!

What can I do?			
1 I can count on and back in 1s, 10s and 100s from any number up to 1000.			
2 I can explain why a number is an odd or an even number.			
3 I can read and write 3-digit numbers and show what each digit stands for.			
4 I can explain the use of zero (0) as a placeholder in a 3-digit number.			
5 I can make a good estimate of the number of objects in a group.			

I need more help with:

_____

_____

_____

Unit 2 Addition and subtraction

Can you remember?

Write the missing number each time.

a 14 + ⬚ = 20   **b** 20 − 6 = ⬚   **c** ⬚ + 10 = 20

d ⬚ − 10 = 10   **e** 20 = ⬚ + 17   **f** 17 = ⬚ − 3

Complements of 100

1 Each bar totals 100. Write the missing number each time.

a | 25 |

b | 55 |

c | 85 |

d | 80 |

e | 65 |

2 Rearrange these digits to total 100 in 4 different ways:

| 1 | 2 | 3 | 4 | 5 | 6 | 7 | 8 |

⬚⬚ + ⬚⬚ = 100   ⬚⬚ + ⬚⬚ = 100

⬚⬚ + ⬚⬚ = 100   ⬚⬚ + ⬚⬚ = 100

Adding in a different order

1 Change the order to make each calculation easier.

a 8 + 11 + 2

⬚ + ⬚ + ⬚ = ⬚

b 11 + 7 + 9

⬚ + ⬚ + ⬚ = ⬚

c 25 + 51 + 5

⬚ + ⬚ + ⬚ = ⬚

d 18 + 80 + 2

⬚ + ⬚ + ⬚ = ⬚

2 Join the numbers you will add first, as shown.
Then write and complete the calculation.

14 + 7 + 6 = ?

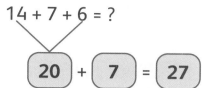

20 + 7 = 27

a   38 + 75 + 25 = ?

◯ + ◯ = ◯

b   22 + 6 + 22 = ?

◯ + ◯ = ◯

c   52 + 19 + 18 = ?

◯ + ◯ = ◯

Adding multiples of 100

1 The numbers in the square and the triangle add up to the number in the circle. Fill in the missing numbers.

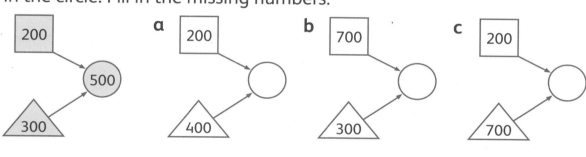

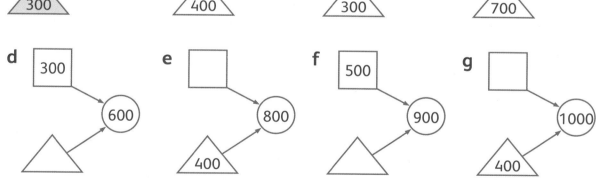

2 To work out the number in a circle, add the 2 numbers below it.
Fill in the missing numbers.

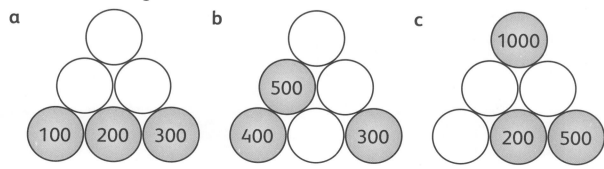

Adding pairs of 2-digit numbers

 1 Write an addition to work out the total length each time.
Complete **a** and **b**. Then do **c** and **d** on your own.

a | 26 cm | 15 cm |

26 cm + 15 cm

26 cm + 10 cm = 36 cm

36 cm + 5 cm = ☐ cm

b | 26 cm | 35 cm |

26 cm + 35 cm

26 cm + 30 cm = ☐ cm

☐ cm + 5 cm = ☐ cm

c | 56 cm | 35 cm |

_____

_____

_____

d | 66 cm | 25 cm |

_____

_____

_____

2 Find the missing numbers in these additions.

a 36 + ☐ = ☐

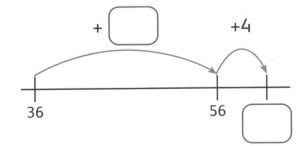

b 47 + ☐ = ☐

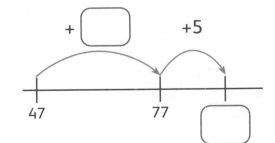

c 65 + 27 = ?

60 + 20 = ☐

☐ + 7 = ☐

☐ + 12 = 92

d 48 + 36 = ?

☐ + 30 = ☐

8 + ☐ = ☐

☐ + 14 = ☐

3 Write an estimate for each calculation. Then work out the answer.
Put a ✓ or ×. If a calculation is ×, write the correction.

Calculation	My estimate	✓ or ×	Correction
11 + 38 = 69	10 + 40 = 50	×	49
22 + 28 = 50			
39 + 42 = 91			
43 + 55 = 98			
67 + 28 = 85			
18 + 53 = 81			
42 + 43 = 85			

Subtracting 2-digit numbers

1 Make 8 subtraction calculations with 2-digit answers.
Use any 4 of the 5 digit cards each time.

1 3 4 6 8

⬜⬜ – ⬜⬜ = ⬜⬜   ⬜⬜ – ⬜⬜ = ⬜⬜

⬜⬜ – ⬜⬜ = ⬜⬜   ⬜⬜ – ⬜⬜ = ⬜⬜

⬜⬜ – ⬜⬜ = ⬜⬜   ⬜⬜ – ⬜⬜ = ⬜⬜

⬜⬜ – ⬜⬜ = ⬜⬜   ⬜⬜ – ⬜⬜ = ⬜⬜

2 Write an estimate for each calculation. Then work out the answer.
Add a ✓ or ×. If a calculation is ×, write the correction.

Calculation	My estimate	✓ or ×	Correction
69 – 38 = 31	70 – 40 = 30	✓	
51 – 28 = 32			
99 – 42 = 47			
87 – 59 = 28			
75 – 26 = 59			
81 – 53 = 18			
63 – 47 = 16			

3 Use a different method each time. The example under each heading shows the methods you can use.

Which method did you like best?
Can you explain why?

Subtracting back	Using a number line	Decomposing or regrouping
a 76 − 25 = 76 − 20 = ☐ ☐ − 5 = ☐	−5 −20 ☐ ☐ 76	76 − 25 = 70 − 20 = ☐ 6 − 5 = ☐ ☐ + ☐ = ☐
b 98 − 34 =		
c 64 − 36 =		
d 83 − 27 =		

Unit 2 Addition and subtraction

Self-check

See how much you know!

	I can do this.
	I can do this, but I need to keep trying.
	I can't do this yet.

What can I do?	😉	😐	🙁
1 I can explain and demonstrate that 9 − 3 = 6, whereas 3 − 9 ≠ 6.			
2 I can show that when numbers are added, their order can be changed but the total does not change.			
3 I can identify complements of 100 from sets of numbers.			
4 I can add pairs of multiples of 100.			
5 I can estimate to check the answers to addition and subtraction calculations.			
6 I can add pairs of 2-digit numbers.			
7 I can take away a 2-digit number from a 2-digit number.			

I need more help with:

_____

_____

_____

Can you remember?

Write the correct mathematical name for each shape.

a _____

b _____

c _____

d _____

e _____

f _____

Shapes around us

1 Label one of each shape you see in these patterns.
Draw a line pointing to the shape and write its name.

a

b

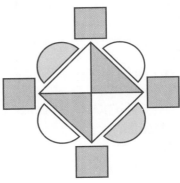

2 Viti and Jack reused materials
to build this rocket.
Write the 3D shapes you see.

_____
_____
_____
_____

3 The diagram shows a construction bridge.
List the 2D shapes you see.
Hint: Look for shapes in other shapes.

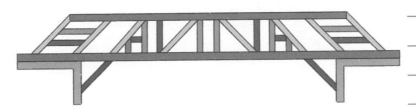

_____
_____
_____
_____

3D shapes and their properties

 1 Write the name of each shape. Now compare the pairs of shapes. Write what is the same and what is different about them.

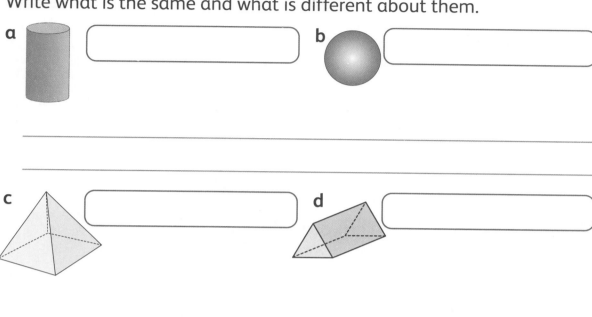

a [   ]

b [   ]

_____

_____

c [   ]

d [   ]

_____

_____

 2 Use the properties to draw each 3D shape. Write the name of the shape. **Hint:** Make each shape with construction materials.

a	3 rectangular faces, 2 triangular faces, 6 vertices	b	8 vertices, all faces the same shape, 12 edges
c	1 square face, 4 triangular faces, 5 vertices, 8 edges	d	8 vertices, 12 edges, 6 rectangular faces

15

3 Complete each shape. Write the number of vertices and edges.

a

Vertices: _____
Edges: _____

b

Vertices: _____
Edges: _____

c

Vertices: _____
Edges: _____

d

Vertices: _____
Edges: _____

4 Look at the shape on the right.

a How many faces can you see? ⬭

How many faces are hidden? ⬭

b Use blocks or other maths equipment to make it.

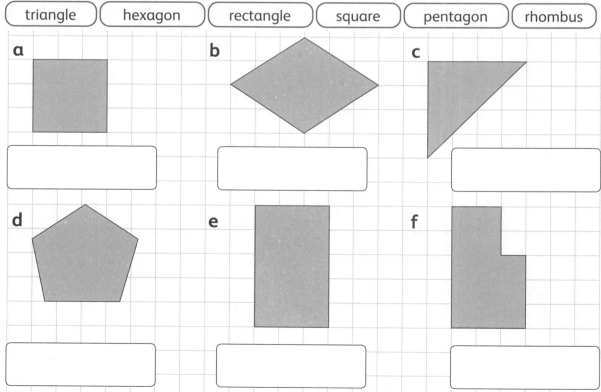

2D shapes and their properties

1 Write the name of each shape in the box under it.

triangle   hexagon   rectangle   square   pentagon   rhombus

a

b

c

d

e

f

2 Complete these sentences.

a A _____ has 4 right angles.

b A _____ has 3 vertices.

c A kite has _____.

d A hexagon has _____ more sides than a rectangle.

e A rectangle has the same number of right angles as a _____.

f An _____ has the same number of sides as a spider has legs.

3 Write sentences to compare each pair of shapes.

Shapes	What is the same and what is different about the shapes?
a	
b	
c	
d	
e	

Regular and irregular polygons

 1 Draw lines to match each regular shape to an irregular shape with the same number of sides.

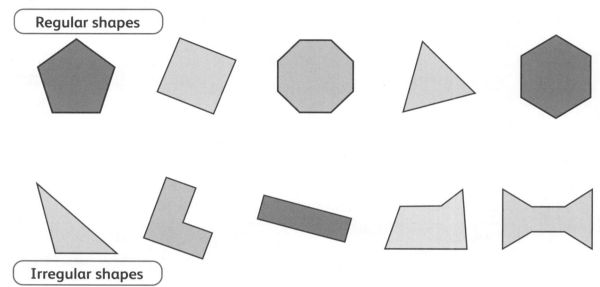

Regular shapes

Irregular shapes

 2 Complete the table.

Shape	Shape name	Regular or irregular? How do you know?
	_____	This shape is _____ because _____ _____
	_____	This shape is _____ because _____ _____
	_____	This shape is _____ because _____ _____
	_____	This shape is _____ because _____ _____

Unit 3 Shapes and angles

Self-check

See how much you know!

 I can do this.

 I can do this, but I need to keep trying.

 I can't do this yet.

What can I do?			
1 I can classify 2D shapes and talk about their properties.			
2 I can sort 3D shapes and talk about their properties.			
3 I can identify regular and irregular polygons.			
4 I know the names of 2D and 3D shapes.			
5 I can compare shapes and say what is the same and what is different about them.			
6 I can create 3D shapes from drawings of the shapes.			

I need more help with:

_____

_____

_____

Unit 4 Statistical methods and chance

Can you remember?

The tally chart shows how many people visited a museum on one day. Fill in the total for each age group.

Age	Tally	Frequency
Under 10	𝍸 𝍸 𝍸 𝍸 𝍸 𝍸	30
Between 10 and 20	𝍸 𝍸 𝍸 𝍸 𝍸	
Between 20 and 30	𝍸 𝍸 𝍸 𝍸 𝍸 𝍸 I	
Between 30 and 40	𝍸 𝍸 𝍸 𝍸 𝍸 𝍸 𝍸 𝍸 I	
Between 40 and 50	𝍸 𝍸 𝍸 𝍸 𝍸 IIII	
Over 50	𝍸 𝍸 I	

Venn diagrams and Carroll diagrams

1 Draw a shape to go in each section of these Venn diagrams.

a

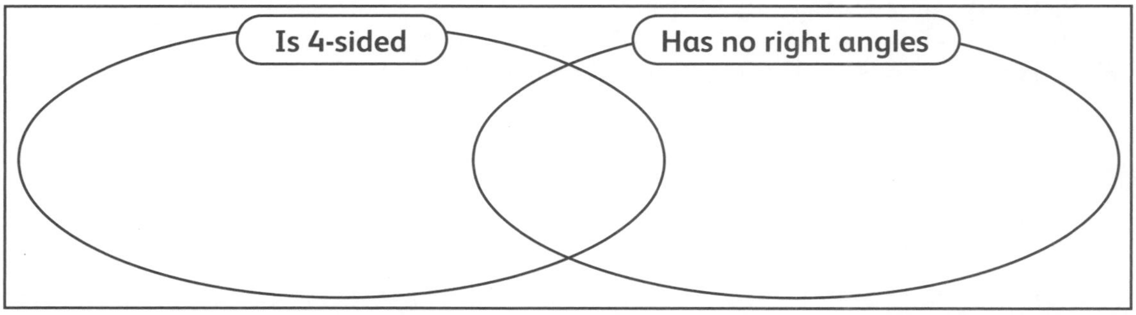
Is 4-sided
Has no right angles

b

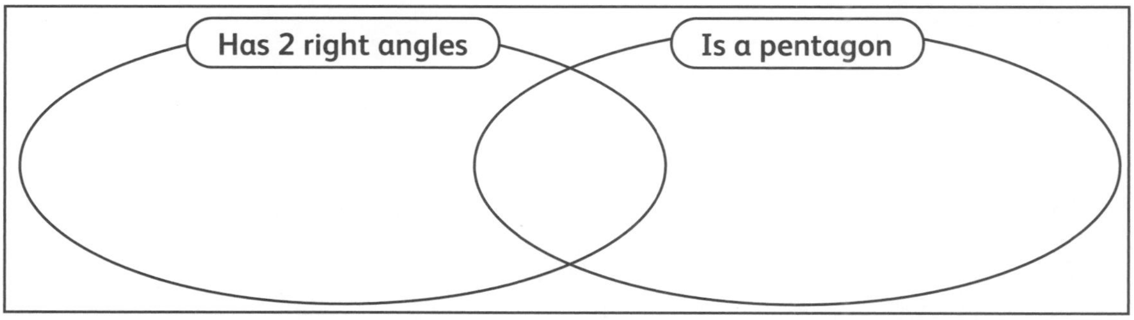
Has 2 right angles
Is a pentagon

2 Write 2 numbers in each section of this Carroll diagram.

	Greater than 10	Not greater than 10
Odd		
Even		

Pictograms and bar charts

1 Jack asked some learners at school to name their favourite colour. The results are in this table.

Favourite colour	Number of learners
Yellow	11
Blue	8
Red	15
Green	9
Pink	3
Purple	8

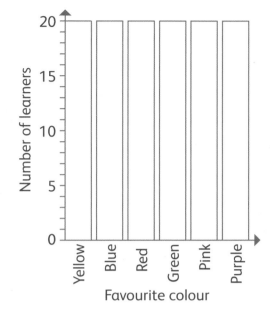

a Show the information in a bar chart.

b Now use the information to draw a pictogram.

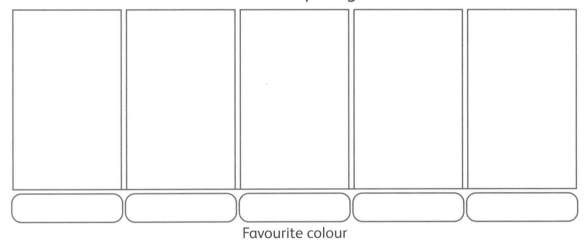

Favourite colour

Key	🎨 represents 2 learners

2 **a** How many learners were there in total?

b How many learners chose blue?

c What was the most popular colour?

d What was the least popular colour?

e How many more learners chose yellow than green?

f How many learners chose red or blue?

Tally charts

1 The soccer score for a game was 2–1.
In total, the players scored 3 goals in that match.
This score card shows the scores for all matches played on the weekend.

2–1	0–0	3–2	1–1	2–0
0–3	1–3	2–0	2–2	1–2
3–1	1–1	3–0	3–1	2–0
1–3	0–1	0–2	2–1	1–2
1–0	2–1	1–1	1–4	0–1

a Complete the tally chart. Show how many matches had each total.

Goals scored	Tally	Number of matches
0 goals		
1 goal		
2 goals		
3 goals		
4 goals		
5 goals		

b What do you notice about the number of goals scored?

_____

_____

c Write 3 questions you could ask about the tally chart.

_____

_____

_____

Self-check

 I can do this.

 I can do this, but I need to keep trying.

 I can't do this yet.

See how much you know!

What can I do?			
1 I can organise information into a list, a table and a chart.			
2 I can record data using a tally chart.			
3 I can answer questions about a pictogram and a bar chart.			
4 I can sort objects and shapes on a Carroll diagram.			
5 I can answer questions about Venn diagrams and Carroll diagrams.			
6 I can interpret and explain the data presented in tables, bar charts and pictograms.			

I need more help with:

_____

_____

_____

Can you remember?

Fill in the missing numbers.

Double 6 is ▢   Double 9 is ▢

Double 10 is ▢   Double 12 is ▢

Double ▢ is 14   Double ▢ is 16

Double ▢ is 22   Double ▢ is 26

The relationship between multiplication and division

1 Draw lines to join the related multiplication and division facts.

$12 \div 2 = 6$ $5 \times 7 = 35$ $10 \times 10 = 100$ $2 \times 8 = 16$

$16 \div 2 = 8$ $2 \times 6 = 12$ $35 \div 5 = 7$ $100 \div 10 = 10$

2 Write a multiplication and a division to go with each array.

a ▢ × ▢ = ▢
▢ ÷ ▢ = ▢

b 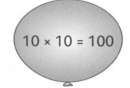 ▢ × ▢ = ▢
▢ ÷ ▢ = ▢

c  ▢ × ▢ = ▢
▢ ÷ ▢ = ▢

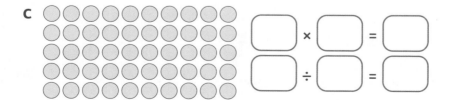

3 Fill in the missing information.

Multiples of 2, 5 and 10

1
a Colour the multiples of **2** in yellow, and the multiples of **5** in blue. If a number is both, make it half yellow and half blue.

1	2	3	4	5	6	7	8	9	10
11	12	13	14	15	16	17	18	19	20
21	22	23	24	25	26	27	28	29	30
31	32	33	34	35	36	37	38	39	40
41	42	43	44	45	46	47	48	49	50

b What do you notice?

_____

2
a Put these numbers in the correct sections of the Carroll diagrams. A number can be used more than once.

(24) (30) (75) (19) (68) (150) (99) (105)

	Multiple of 2	Not a multiple of 2
Multiple of 10		
Not a multiple of 10		

	Multiple of 5	Not a multiple of 5
Multiple of 10		
Not a multiple of 10		

b Explain why a section is left empty each time.

_____

_____

The multiplication tables of 2, 4 and 8

1 Complete these multiplication grids.

a

×	6	9	
2			
4		40	
8			

b

×	2	4	8
3		12	
	10		
7			

2 How many sides in total?

a 6 squares → ☐ sides   b 6 octagons → ☐ sides

c 9 squares → ☐ sides   d 9 octagons → ☐ sides

3 True or false? Write **T** or **F**.

a 4 × 6 = 6 × 4 ☐   b 8 × 7 = 8 × 8 ☐

c 2 × 9 = 9 × 2 ☐   d 4 × 9 = 9 × 3 ☐

The multiplication tables of 3, 6 and 9

1 Write the multiplication fact for each array. Then shade squares in another colour to show the second multiplication fact.

a

3 × ☐ = ☐

6 × ☐ = ☐

b

3 × ☐ = ☐

9 × 5 = ☐

2 Complete these multiplication grids.

a

×	4		7
3		18	
6			
9			

b

×	3	6	9
3			
		12	
7			

3 Maris has 4 marbles.
Jack has 3 times as many marbles as Maris.
Viti has 6 times as many marbles as Maris.

a How many marbles do Jack and Viti have altogether?

b Jack and Viti put their marbles in groups of 9.
How many groups can they make between them?

4 True or false? Write **T** or **F**.

a 3 × 5 = 3 × 3

b 6 × 8 = 8 × 6

c 9 × 5 = 5 × 7

d 6 × 4 = 4 × 6

Multiplying numbers

1 Draw lines to mark the pair of numbers you will multiply first.
Then complete the multiplication.

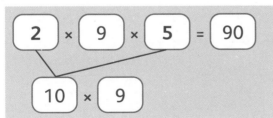

2 × 9 × 5 = 90
10 × 9

a  7 × 2 × 5 =

☐ × ☐

b 6 × 5 × 2 =

☐ × ☐

c 9 × 3 × 3 =

☐ × ☐

2 Complete these.

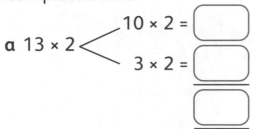

a 13 × 2 ⟨ 10 × 2 = ☐
              3 × 2 = ☐ +
              ☐

b 17 × 4 ⟨ 10 × 4 = ☐
              7 × 4 = ☐ +
              ☐

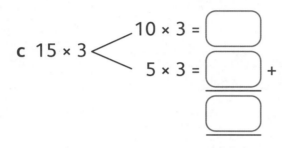

c 15 × 3 ⟨ 10 × 3 = ☐
              5 × 3 = ☐ +
              ☐

d 5 × 18 ⟨ 5 × 10 = ☐
              5 × 8 = ☐ +
              ☐

3 A pet shop sells sawdust in 2 kg and 4 kg bags.
There are 16 of the 2 kg bags of sawdust.
There are also 16 of the 4 kg bags of sawdust.

a Calculate the total mass of the 2 kg bags.

_____

_____

_____

b Calculate the total mass of the 4 kg bags.

_____

_____

_____

c There is a relationship between the total masses each time.
What is the relationship? Explain why this happens.

_____

_____

_____

Self-check

 I can do this.

 I can do this, but I need to keep trying.

 I can't do this yet.

See how much you know!

What can I do?			
1 I can understand the inverse relationship between multiplication and division.			
2 I can show that when numbers are multiplied their order can be changed without the answer changing.			
3 I can model decomposing and regrouping with numbers to 20 to multiply.			
4 I can use known multiplication tables facts for the 2, 3, 4, 5 and 10 times tables to help me to recall other times tables facts.			
5 I can find multiples of 2, 5 and 10 from a set of numbers past the tenth multiple.			

I need more help with:

_____

_____

_____

Can you remember?

Draw the hands on the analogue clocks to show the digital times.

a

09:35

b

02:05

c

12:50

Time

1 Match the activity to the correct unit of time.

hours   minutes   years

2 Draw the hands to show the time on each clock.

a

10 minutes after 4:30   10 minutes before 4:30

b

5 minutes after 5:45   5 minutes before 5:40

c

20 minutes after 7:40   20 minutes before 7:40

d

20 minutes after 7:45   30 minutes after 7:45

Length

1 Look at each line. First estimate the length and record it. Then measure the line and record it.

a _____

Estimate [ ] cm   Measure [ ] cm

b _____

Estimate [ ] cm   Measure [ ] cm

c _____

Estimate [ ] cm   Measure [ ] cm

d _____

Estimate [ ] cm   Measure [ ] cm

e _____

Estimate [ ] cm   Measure [ ] cm

f _____

Estimate [ ] cm   Measure [ ] cm

2 Convert these measurements.

a 3 m = [ ] cm   **b** 3 km = [ ] m

c 5 m = [ ] cm   **d** 5 km = [ ] m

e 7 m = [ ] cm   **f** [ ] m = 800 cm

g [ ] km = 8000 m   **h** [ ] km = 1000 m

i [ ] m = 100 cm   **j** [ ] cm = 2 m

3 X marks the spot of the ant's nest. Draw 3 different routes for the ant to get back to the nest. Each route must be exactly 20 cm long. Use a different colour each time. Stay on the grid lines.

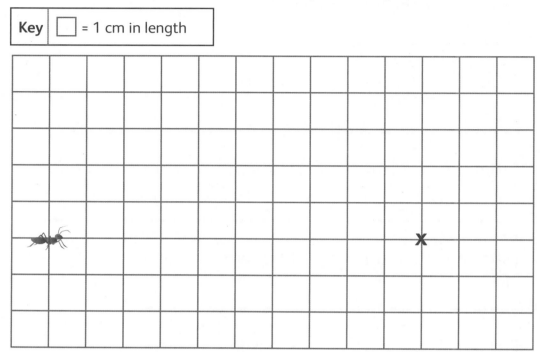

Key ☐ = 1 cm in length

4 Fill in the missing numbers on this ruler.

5 Which unit would you use to measure the length of each object?

millimetres (mm)   centimetres (cm)   metres (m)   kilometres (km)

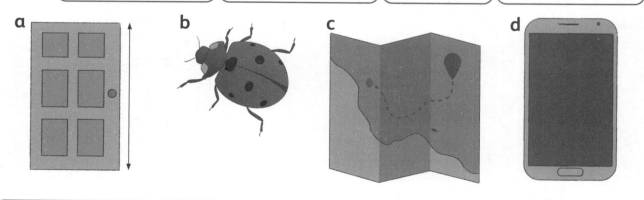

a   b   c   d

Unit 6 Time and measurement

Self-check

See how much you know!

	I can do this.
	I can do this, but I need to keep trying.
	I can't do this yet.

What can I do?	☺	😐	☹
1 I can identify which units of time to use for different activities.			
2 I can say the time accurately as minutes past the hour and write it in digital notation.			
3 I can estimate lengths in centimetres (cm), metres (m) and kilometres (km) before measuring.			
4 I can convert between mm, cm, m and km.			
5 I can choose suitable units to estimate and measure length.			
6 I can say what one division on a scale is worth.			

I need more help with:

_____

_____

_____

Can you remember?

Use each number card once to complete the calculations.

100	200	300	400
400	500	600	700

$\boxed{} + \boxed{} = 500$ $\boxed{} + \boxed{} = 800$

$\boxed{} + \boxed{} = 900$ $\boxed{} + \boxed{} = 1000$

Calculations with missing numbers

1 Fill in the missing numbers to complete the puzzles.

a

3	+		=	10
+	▓	−	▓	▓
	−		=	
=	▓	=		
9	▓	2		

b

	+		=	17
+	▓	−	▓	▓
12	−		=	
=	▓	=		
18	▓	4		

2 Find the value of each shape.

a ⭐ + 7 = 10   ⭐ = _____

b ⬮ + ⬮ = 12   ⬮ = _____

c $40 − 🙂 = $30   🙂 = $ _____

d ◆ − $5 = $4   ◆ = $ _____

Adding multiples of 10

1 Complete these bar models.

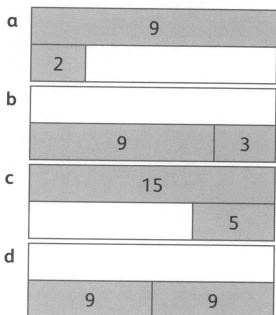

a | 9 | | 90 |
2 | | 20 |

b | | | |
9 | 3 | 90 | 30 |

c | 15 | | 150 |
| 5 | 50 | |

d | | | |
9 | 9 | 90 | 90 |

2 Draw lines to match each calculation with the same total.
Then write a calculation in the empty box to match the
leftover calculation.

160 + 40   20 + 140

90 + 50   80 + 90

30 + 150   70 + 70

120 + 40   30 + 170

20 + 110   [ ]

150 + 20   60 + 70

Adding pairs of 2-digit and 3-digit numbers

The grid on the right shows
the numbers **173** and **12**.
Add the numbers:
173 + 12 = 185.

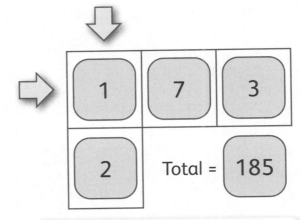

Fill in the grids below, using
the digits 2, 3, 4, 5 and 6.
Each grid must have
a different total.

Could you make more additions
with the digits 2, 3, 4, 5 and 6?
How do you know?
Write your answer in part **e**.

a

Total =

b

Total =

c

Total =

d

Total =

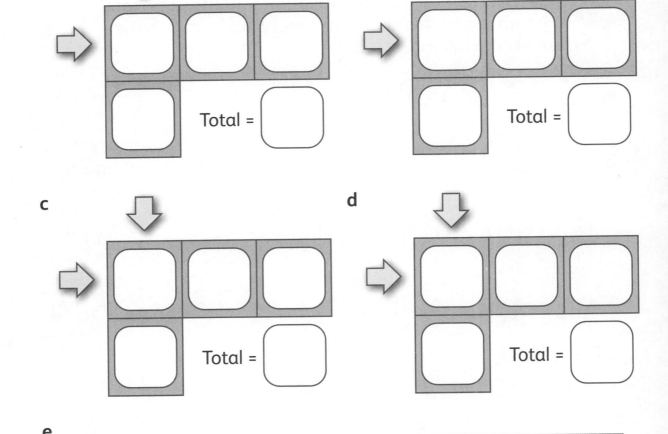

e _____

_____

2 Complete these. First estimate. Then calculate.

Calculation	Estimate	Use an adding on or a regrouping method each time.
a 142 + 27		
b 325 + 48		
c 481 + 53		

3 Find the missing digits.

a

100s	10s	1s
2	3	( )
+	4	2
2	( )	7

b

100s	10s	1s
4	3	( )
+	4	9
( )	8	4

c

100s	10s	1s
5	6	( )
+	( )	5
6	3	9

d

100s	10s	1s
5	( )	( )
+	5	2
5	5	8

e

100s	10s	1s
4	( )	1
+	3	( )
4	9	9

f

100s	10s	1s
2	( )	4
+	7	( )
3	5	7

Subtracting 2-digit numbers from 3-digit numbers

1 Complete these. First estimate. Then calculate.

	Calculation	Estimate	Use a subtracting back or a regrouping method each time.
a	157 – 36		
b	363 – 25		
c	427 – 64		

2 Find the missing digits.

a
100s	10s	1s
4	☐	9
–	3	☐
4	4	3

b
100s	10s	1s
5	7	☐
– ☐		8
5	3	4

c
100s	10s	1s
3	☐	4
–	7	☐
2	6	1

d
100s	10s	1s
2	7	☐
–	4	2
2	☐	5

e
100s	10s	1s
3	☐	☐
–	4	2
3	5	6

f
100s	10s	1s
4	8	☐
–	4	7
☐	3	7

3 Solve these problems. Calculate them in your own way.

Problem	My working
a 468 people are watching a match. 43 people leave at half time. How many people are left?	
b A loaf of bread costs 175 cents. A carton of milk costs 36 cents less. How much is a carton of milk?	
c A length of wood is 348 cm. Another length is 55 cm shorter. How long is the shorter length?	

Working with money

1 Order each set of money from smallest to largest.

a

21c	12c	221c	120c	210c	121c	201c

b

29c	1c	209c	299c	199c	200c	92c

c

$1.20	102c	21c	$1.01	111c	11c

d

$2.02	220c	$2	$2.22	$2.01	212c

2 Write 5 ways to spend $1. Choose 2 or more items each time.

_____

_____

_____

_____

_____

19c 45c 10c 51c 9c

5c 50c 11c 49c 15c

40c 21c 39c 25c 30c

33c 29c 35c 20c 41c

3 Each person starts with $1, but then spends some of it.
Work out the missing change or amount spent.

	Amount spent	Change
a	25c, 25c	◯
b	25c, 10c	◯◯◯
c	50c, 25c, 10c	◯◯
d	50c + ◯ + ◯	30c
e	50c + ◯ + 5c	20c + ◯

4 **a** Maris buys 2 different items. She gets 25c change from $1.
How much could each item be? ☐ cents and ☐ cents

b Jack buys 2 identical items. He gets 32c change from $1.
How much is each item? ☐ cents

c David buys some items and gets $2 change from a $10 note.
Complete these to show what the items could have cost.

$☐ and $☐ $☐ , and $☐

$☐ , $☐ , $☐ and $☐

Self-check

 I can do this.

 I can do this, but I need to keep trying.

 I can't do this yet.

See how much you know!

What can I do?			
1 I can work out the value of an unknown quantity in an addition calculation and in a subtraction calculation.			
2 I can identify complements of 100 from sets of numbers.			
3 I can add pairs of multiples of 10.			
4 I can estimate to check the answers to addition and subtraction calculations.			
5 I can add pairs of 2-digit and 3-digit numbers.			
6 I can take away a 2-digit number from a 3-digit number.			
7 I can use money notation with a decimal point, knowing that, for example, $2.50 means 2 dollars and 50 cents.			
8 I can use notes and coins to pay for items and work out the change.			

I need more help with:

_____

_____

_____

Can you remember?

Draw place value equipment to show each number.

213	231	302

Multiplying by 10

1 Multiply each number by 10 and write it in the next row. Complete the calculation.

a 43 × 10 = ☐

100s	10s	1s
	4	3

b 23 × 10 = ☐

100s	10s	1s
	2	3

c 93 × 10 = ☐

100s	10s	1s
	9	3

2 Match each multiplication to the correct answer.

60 × 10	630
40 × 10	360
4 × 10	60
6 × 10	400
63 × 10	40
36 × 10	600

3 Viti has made a mistake. What is it? Draw a diagram to help Viti understand her mistake.

36 × 10 = 306

4 Complete these calculations.

a 50 + 100 = ☐ × 10

b 200 − ☐ = 19 × 10

c 50 + 50 + 50 + 50 = ☐ × 10

d 10 × ☐ = 999 − 90 − 9

Comparing and ordering 3-digit numbers

1 Order these numbers.
Use the number line to help.

[ ] < [ ] < [ ] < [ ]

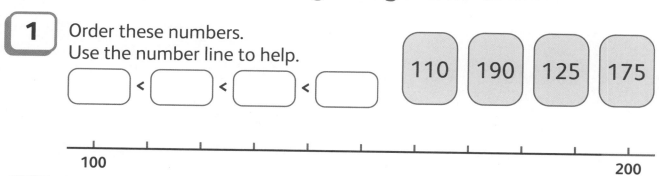

110 190 125 175

100   200

2 Use this number line to complete each statement using **<** or **>**.

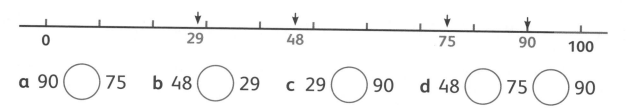

0   29   48   75   90 100

a 90 ◯ 75   **b** 48 ◯ 29   **c** 29 ◯ 90   **d** 48 ◯ 75 ◯ 90

3 Spin a spinner. Write the score as a digit in one of the boxes.
Keep going until you have filled every digit box.
Check if each statement is true or not.
The aim is for both statements to be true at the end.

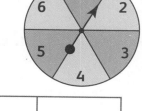

Game 1

			>			
			<			

Now try again for the different games below.
Talk about the tactics you will use.

Game 2

		>		
		<		

Game 3

			>			
			<			

4 Complete each statement using either <, > or =.

a 3 × 10 ◯ 25

b 45 ◯ 10 × 4

c 130 + 10 ◯ 145

d 310 + 30 ◯ 345

e 200 ◯ 260 − 50

f 200 ◯ 250 − 60

Rounding to the nearest 100

1 Draw an arrow to show each number on the number line.

| 390 | 301 | 310 | 351 | 349 | 309 |

300   400

Round each number to the nearest 100.

390 rounds to ▢   301 rounds to ▢   310 rounds to ▢

351 rounds to ▢   349 rounds to ▢   309 rounds to ▢

2 Use these digit cards to make 3-digit numbers.

2   5   9   3   0

Rounds to 200	Rounds to 300	Rounds to 400	Rounds to 500	Rounds to 600
◯ ◯ ◯	◯ ◯ ◯	◯ ◯ ◯	◯ ◯ ◯	◯ ◯ ◯
◯ ◯ ◯	◯ ◯ ◯	◯ ◯ ◯	◯ ◯ ◯	◯ ◯ ◯
◯ ◯ ◯	◯ ◯ ◯	◯ ◯ ◯	◯ ◯ ◯	◯ ◯ ◯

3 Who is right? Annay or David? Explain using words or a diagram.

I agree with ▢ because

_____

_____

45 rounds to 0.

No. You cannot round to 0. 45 rounds to 100, as that is the first hundred.

Number patterns

1 Continue each counting pattern.

a 10, 20, 30, ⬭ , ⬭ , ⬭ , ⬭ , ⬭ , ⬭ , 100

b 11, 111, 211, ⬭ , ⬭ , ⬭ , ⬭ , ⬭ , ⬭

c 909, ⬭ , ⬭ , ⬭ , ⬭ , ⬭ , ⬭ , ⬭ , ⬭ , 9

2 Invent your own counting patterns. Follow the rule each time.
Start with a number between 500 and 600 for each pattern.

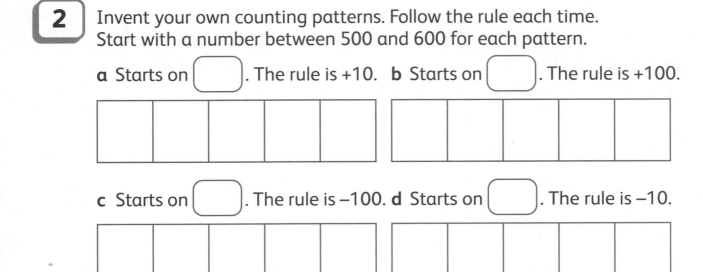

a Starts on ⬭ . The rule is +10. b Starts on ⬭ . The rule is +100.

c Starts on ⬭ . The rule is −100. d Starts on ⬭ . The rule is −10.

3 Draw the next stages of this growing pattern.

Fill in how many dots. Write the rule: _____

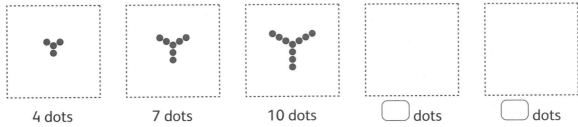

4 dots   7 dots   10 dots   ⬭ dots   ⬭ dots

4 Invent your own growing pattern. Then draw it.
Write the rule: _____

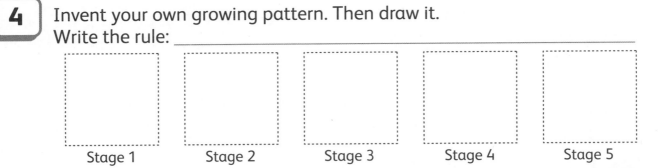

Stage 1   Stage 2   Stage 3   Stage 4   Stage 5

Unit 8 Patterns, place value and rounding

Self-check

 I can do this.

 I can do this, but I need to keep trying.

 I can't do this yet.

See how much you know!

What can I do?	😄	😐	😟
1 I can explain the result of multiplying a number by 10 and demonstrate using a place value chart.			
2 I know what each digit means in 3-digit numbers.			
3 I can compare 3-digit numbers using the symbols >, < and =.			
4 I can order a set of 3-digit numbers on a number line.			
5 I can round 3-digit numbers to the nearest 10 or 100.			
6 I can identify the rule for sequences of numbers.			
7 I can continue a pattern of cubes that increases by 3 each time and explain the rule.			

I need more help with:

_____

_____

_____

Unit 9   Multiplication and division

Can you remember?

Complete these.

Half of 18 is [ ].   Half of 24 is [ ].   Half of 32 is [ ].

Half of [ ] is 11.   Half of [ ] is 13.   Half of [ ] is 15.

Learning multiplication tables

1 Use what you know about doubling to help find the missing numbers.

a $4 \times 3 = 12$, so $8 \times 3 = $ [ ]   b $2 \times 5 = 10$, so $4 \times 5 = $ [ ]

c $3 \times 7 = $ [ ], so $6 \times 7 = $ [ ]   d $3 \times 10 = $ [ ], so $6 \times 10 = $ [ ]

2 Use 2 pencil crayon colours each time. Show that:

a 9×3 is equal to triple 3×3 b 9×4 is equal to triple 3×4

$3 \times 3 = $ [ ] and $9 \times 3 = $ [ ] $3 \times 4 = $ [ ] and $9 \times 4 = $ [ ]

3 Use the digit cards 1 to 5.
Make different 2-digit numbers that are:

a in both the 2× and the 3× table

_____

b in both the 5× and the 4× table

_____

c in both the 5× and the 3× table

_____

Using multiplication and division facts

1 How many groups of stars are in each box?

Groups of 2 stars

There are ☐ groups because

☐ × ☐ = ☐

Groups of 3 stars

There are ☐ groups because

☐ × ☐ = ☐

2 Write a division calculation to solve each problem.
Also write the multiplication fact that can help you.

  a Viti shares 18 sweets equally into 3 bags.
  How many sweets are in each bag?

  ☐ ÷ ☐ = ☐ because ☐ × ☐ = ☐

  b There are 32 legs in a field of sheep.
  How many sheep altogether?

  ☐ ÷ ☐ = ☐ because ☐ × ☐ = ☐

  c A group of 4 children make $28 by selling lemonade.
  They share the money equally.
  How much does each child get?

  ☐ ÷ ☐ = ☐ because ☐ × ☐ = ☐

Multiplying 2-digit numbers

1 Complete these. Show your working under each multiplication.

a 18 × 2 = ⬜	b 2 × 12 = ⬜	c 18 × 4 = ⬜	d 12 × 4 = ⬜

2 Solve these estimation problems. Show any estimates you make.

a Lollies cost 18 cents each. Is 80 cents enough to buy 4 lollies?

_____

_____

b The mass of a rubber is 16 g. Will the mass of 3 rubbers be more than 60 g?

_____

_____

c A tank holds 60 litres of water. A smaller container holds 17 litres of water. Is the tank large enough to hold the water from 5 smaller containers?

_____

_____

3 Write matching multiplications and solve these.

		My working
a	10 1 1 1 10 1 1 1 10 1 1 1 10 1 1 1	⬜ × ⬜ = ⬜
b	10 1 1 1 1 1 10 1 1 1 1 1 10 1 1 1 1 1	⬜ × ⬜ = ⬜
c	10 1 1 1 1 1 1 1 10 1 1 1 1 1 1 1 10 1 1 1 1 1 1 1 10 1 1 1 1 1 1 1 10 1 1 1 1 1 1 1	⬜ × ⬜ = ⬜

Dividing 2-digit numbers

1 Complete these. Show your working under each division.

a 68 ÷ 2 = ☐	**b** 76 ÷ 2 = ☐	**c** 84 ÷ 2 = ☐	**d** 52 ÷ 2 = ☐

2 Complete the division sentences to match these arrays.

a

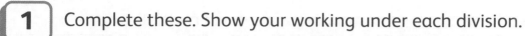

☐ ÷ 4 = ☐

b

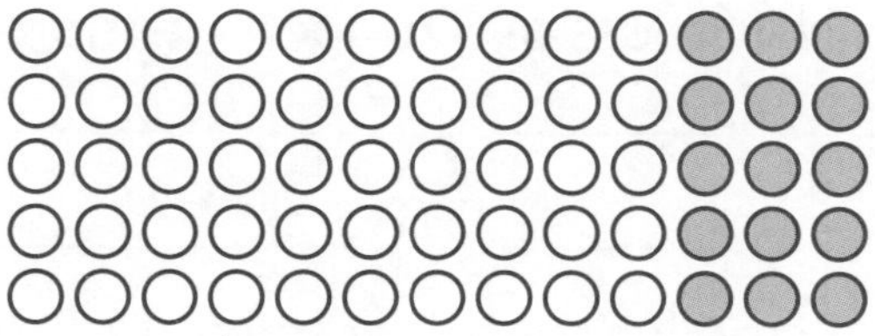

☐ ÷ 5 = ☐

c

☐ ÷ 3 = ☐

3 Use decomposing or regrouping to solve these.

a 39 ÷ 3 = ? 30 ÷ 3 = ☐

9 ÷ 3 = ☐ +

☐

b 56 ÷ 4 = ? 40 ÷ 4 = ☐

16 ÷ 4 = ☐ +

☐

c 85 ÷ 5 = ? 50 ÷ 5 = ☐

☐ ÷ 5 = ☐ +

☐

d 51 ÷ 3 = ? 30 ÷ 3 = ☐

☐ ÷ 3 = ☐ +

☐

Solve these problems. Show any estimates that you make.

Problem	Estimate	My working
a $54 is shared equally between 3 people. How much does each person get? $⬚		
b The total mass of 5 identical boxes is 85 kg. What is the mass of 1 box? ⬚ kg		
c A train travels the same journey 4 times. It travels a total of 84 km. What is the length of each journey? ⬚ km		

Unit 9 Multiplication and division

Self-check

 I can do this.

 I can do this, but I need to keep trying.

 I can't do this yet.

See how much you know!

What can I do?			
1 I can simplify a calculation to multiply by changing the order of the numbers I am multiplying.			
2 I can model decomposition to multiply.			
3 I can recall the 1, 2, 3, 4, 5, 6, 8, 9 and 10 times tables, using known facts to recall others.			
4 I can recall division facts that are related to known multiplication facts.			
5 I can estimate to check the answers to multiplication and division calculations.			
6 I can multiply any 2-digit number by 2, 3, 4 and 5.			
7 I can divide 2-digit numbers by 2, 3, 4 and 5 with no remainders.			

I need more help with:

_____

_____

_____

Can you remember?

Write the digital times to match the analogue clocks.

a

b

c

Time

1 A train timetable

Station	Arrival time
A	11:40
B	11:45
C	11:55
D	12:10
E	12:15
F	12:30

What time does the train arrive at train station A?

What time does the train arrive at train station C?

How many minutes is it between a train arriving at station B and arriving at station D?

Which station does the train reach at 12:15?

Annay gets off the train at 12:10. Which station is he at?

Mass

1 Match the measuring instrument to what you would use to measure the:

mass of 4 potatoes?

height of a sunflower?

length of a pencil?

liquid for a recipe?

2 Mark the mass on each scale.

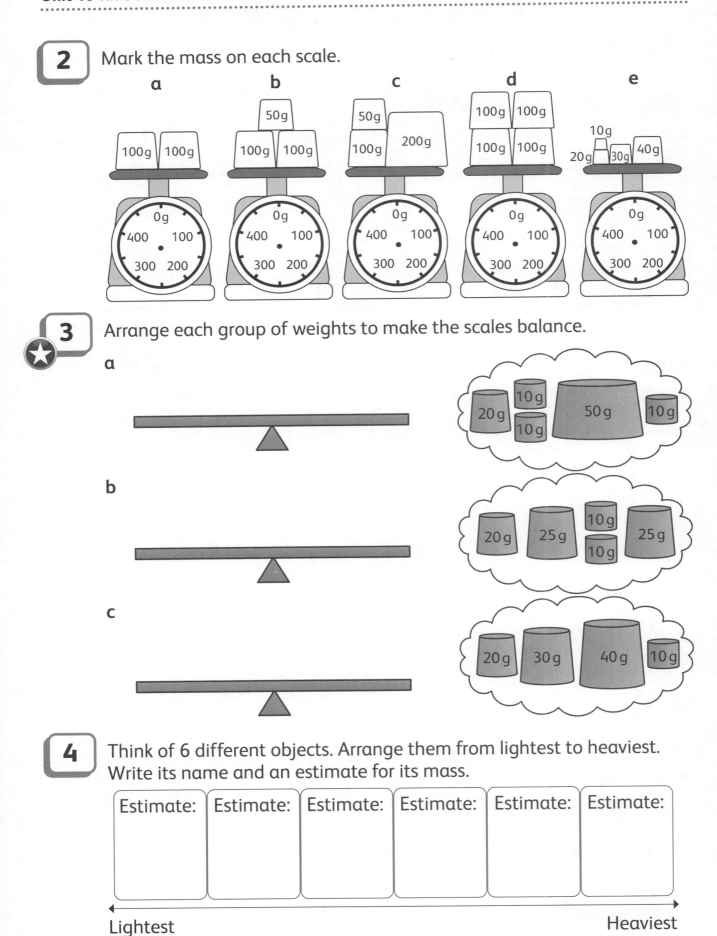

a   b   c   d   e

3 Arrange each group of weights to make the scales balance.

a

b

c

4 Think of 6 different objects. Arrange them from lightest to heaviest.
Write its name and an estimate for its mass.

Estimate:	Estimate:	Estimate:	Estimate:	Estimate:	Estimate:

Lightest   Heaviest

Capacity

1 Draw the level of the amount of liquid given on the jugs.
Each line is 100 ml.

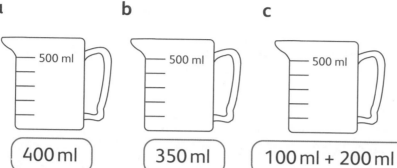

a
| 400 ml |

b
| 350 ml |

c
| 100 ml + 200 ml |

d
| 800 ml + 150 ml |

2 Fill in the missing information.

a [   ] ml = 4ℓ

b [   ] ml = 5ℓ

c [   ] ml = $6\frac{1}{2}$ℓ

d   8 500 ml = [   ] $\frac{1}{2}$ℓ

e   7 500 ml = [   ] ℓ

f [   ] ml = $9\frac{1}{4}$ℓ

3 Share the amount between the pairs of jugs each time.
Draw the level on each jug. They do not need to be equal.

a

| 200 ml |

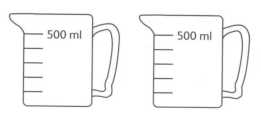

b

| 600 ml |

c
| 450 ml |

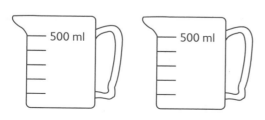

d
| 850 ml |

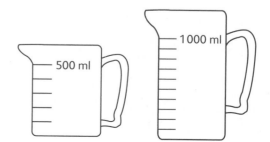

Unit 10　Time and measurement

Self-check

 I can do this.

 I can do this, but I need to keep trying.

 I can't do this yet.

See how much you know!

What can I do?			
1 I can read and use the information shown on a timetable.			
2 I can use scales to measure the mass of objects in kilograms (kg) and grams (g).			
3 I can estimate and measure mass accurately using grams (g) and kilograms (kg).			
4 I can estimate and measure capacity using litres (ℓ) and millilitres (ml).			
5 I can read a scale to the nearest division or half-division.			
6 I can select and use measuring instruments for length, mass and capacity.			

I need more help with:

_____

_____

_____

Unit 11   Shapes and angles

Can you remember?

Tick (✓) all the irregular shapes.

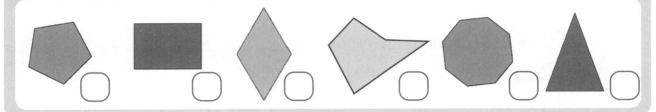

Symmetrical shapes

1 Complete the symmetrical shapes. Then write the name of each.

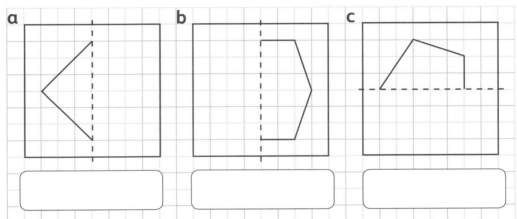

a

b

c

2 Draw each shape in 2 ways.
Then tick (✓) the properties that match each shape.

Name	Drawing A	Drawing B	Even number of sides	Odd number of vertices	Always symmetrical
Rectangle					
Hexagon					
Pentagon					
Rhombus					
Kite					

3 Look at the shapes. Name the ones with the lines of symmetry described below. Say if it is a regular shape or not.

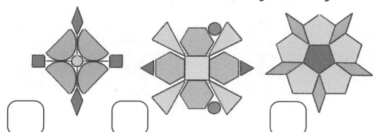

a Only a vertical line of symmetry _____

b Vertical and horizontal lines of symmetry _____

c More than 2 lines of symmetry _____

d No lines of symmetry _____

4 a Look at the 3 patterns below. Tick (✓) if it has only 1 horizontal line of symmetry. Circle (O) if it has only a vertical line of symmetry. Cross (×) if it has a horizontal and a vertical line of symmetry.

b Tick (✓) the buildings that have lines of symmetry.

Reflecting shapes

1 Jack draws some shapes. He reflects them in the mirror lines and draws the reflections. Tick (✓) the correctly reflected shapes.

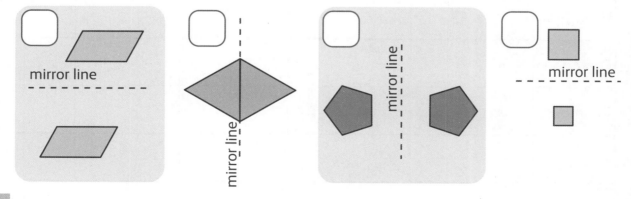

2 Use a mirror to reflect these shapes. Sketch the reflected shape(s).

a

mirror line _ _ _ _ _

b

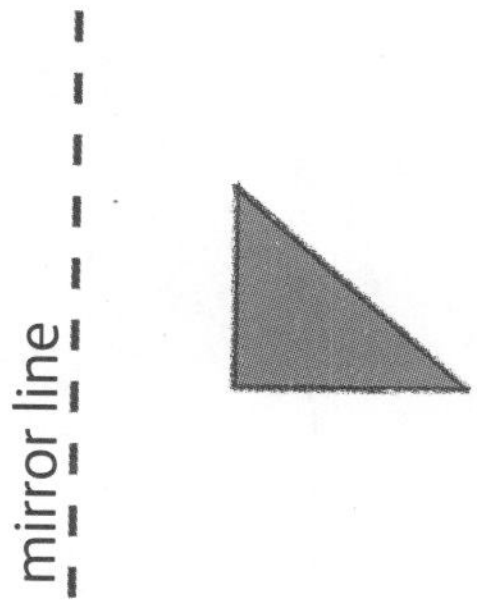

c

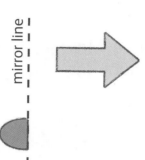

d

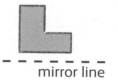

mirror line

Angles

1 Decide whether each angle is **>** or **<** than a right angle (∟).
Tick (✓) a box each time.

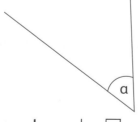

angle a > ∟ ☐
angle a < ∟ ☐

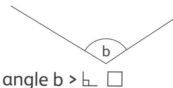

angle b > ∟ ☐
angle b < ∟ ☐

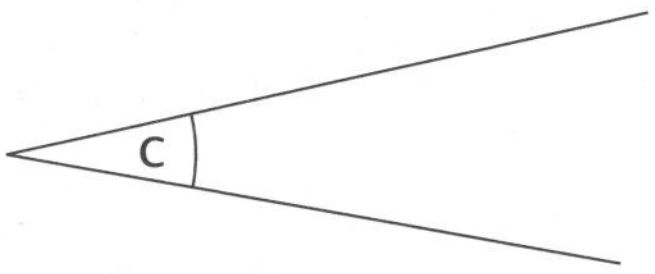

angle c > ∟ ☐
angle c < ∟ ☐

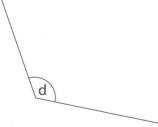

angle d > ∟ ☐
angle d < ∟ ☐

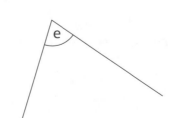

angle e > ∟ ☐
angle e < ∟ ☐

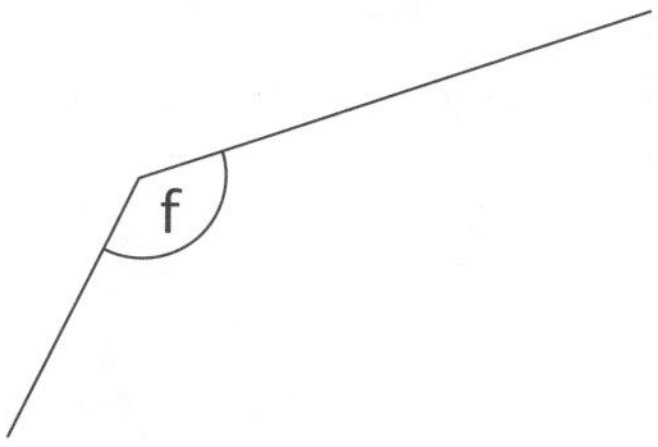

angle f > ∟ ☐
angle f < ∟ ☐

2 Draw a shape that belongs in each section of the Venn diagrams.

a

| Has 4 sides | Has no right angles |

b

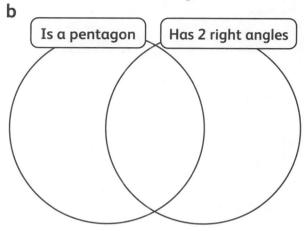

| Is a pentagon | Has 2 right angles |

3 Annay has joined 2 shapes each time. Make a circle (**O**) where you can see that 2 right angles make a straight angle.

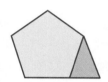  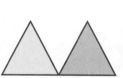

4 Complete these sentences.

a A quarter turn is made up of _____ right angle.

b A half turn is made up of _____ right angles.

c When the minute hand on a clock moves through 2 right angles, it is the same as a _____ turn.

d When the minute hand on a clock moves through 1 right angle, it is the same as a _____ turn.

Position, direction and movement

1 Describe the position of the frog.
Write if it is **above**, **below**, **next to** or **between** the lily pads.

a
_____

b
_____

c
_____

d
_____

2 Follow the instructions. Draw the path you make each time.

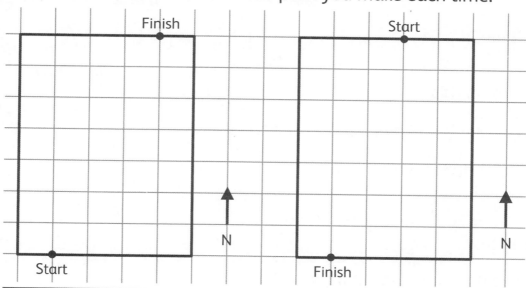

a
Move 2 squares north.
Make a clockwise right-angle turn.
Move 3 squares east.
Make an anticlockwise right-angle turn.
Move 5 squares north.

b
Move 1 square south.
Make an anticlockwise right-angle turn.
Move 1 square east.
Make a clockwise right-angle turn.
Move 3 squares south.
Make a clockwise right-angle turn.
Move 3 squares west.
Make an anticlockwise right-angle turn.
Move 3 squares south.

3 Tick (✓) the diagram that shows all 5 of these directions:

Straight on
Left turn
Straight on
Right turn
Straight on

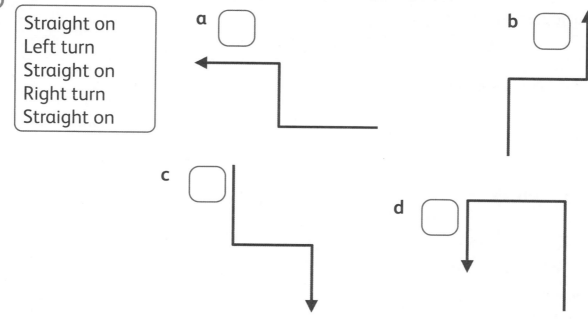

Unit 11 Shapes and angles

Self-check

See how much you know!

 I can do this.

 I can do this, but I need to keep trying.

 I can't do this yet.

What can I do?			
1 I can see if a shape has more than 1 line of symmetry.			
2 I can find shapes with horizontal and vertical lines of symmetry.			
3 I can sketch the reflection of a shape in a horizontal or vertical mirror line, including where the mirror line is the edge of the shape.			
4 I can test whether an angle is equal to, bigger than or smaller than a right angle.			
5 I can understand that a straight line is equivalent to 2 right angles or a half turn.			
6 I can follow and give instructions to make turns and movements on a grid.			

I need more help with:

_____

_____

_____

Can you remember?

a Colour in $\frac{1}{2}$ of each shape.

b Colour in $\frac{1}{4}$ of each shape.

  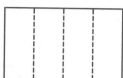

Three quarters

1 Tick (✓) the shapes that show $\frac{3}{4}$.

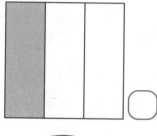 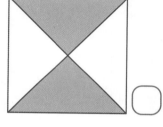 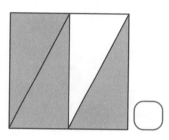

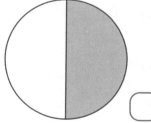 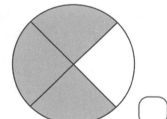 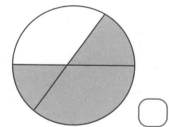

2 Draw a diagram to show each fraction.

$\frac{1}{4}$	$\frac{2}{4}$	$\frac{3}{4}$	$\frac{4}{4}$

3 Circle the dots to find each fraction.

a

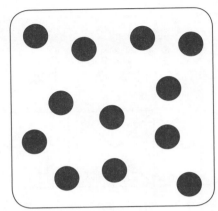

$\frac{1}{4}$ of 12 = ☐

b

$\frac{1}{4}$ of 24 = ☐

c

$\frac{3}{4}$ of 12 = ☐

d

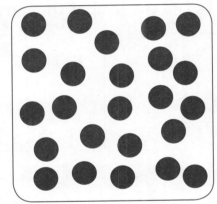

$\frac{3}{4}$ of 24 = ☐

4 Complete these sets of statements.

a $\frac{1}{4}$ of ☐ = 1

$\frac{1}{4}$ of ☐ = 2

$\frac{1}{4}$ of ☐ = 3

$\frac{1}{4}$ of ☐ = 5

b $\frac{3}{4}$ of ☐ = 3

$\frac{3}{4}$ of ☐ = 9

$\frac{3}{4}$ of ☐ = 30

$\frac{3}{4}$ of ☐ = 24

Equal parts of a whole

1 Match each diagram to a fraction card.

| $\frac{1}{3}$ | $\frac{1}{5}$ | $\frac{1}{6}$ | $\frac{1}{10}$ |

2 Shade each diagram to show the fraction.

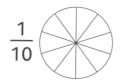  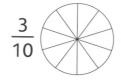  

$\frac{1}{5}$ $\frac{2}{5}$ $\frac{3}{5}$ $\frac{4}{5}$ $\frac{5}{5}$

$\frac{1}{10}$ $\frac{2}{10}$ $\frac{3}{10}$ $\frac{4}{10}$ $\frac{5}{10}$

$\frac{1}{9}$ $\frac{2}{9}$ $\frac{8}{9}$ $\frac{9}{9}$

3 Play with a partner. Take turns to choose a fraction from the grid. Draw a diagram to match the fraction. If you are correct, you win that box. Cover it with a counter. Try to win 3 in a row.

$\frac{1}{3}$	$\frac{1}{4}$	$\frac{1}{5}$	$\frac{1}{6}$	$\frac{1}{7}$
$\frac{2}{3}$	$\frac{2}{5}$	$\frac{2}{7}$	$\frac{2}{8}$	$\frac{2}{2}$
$\frac{6}{6}$	$\frac{3}{3}$	$\frac{5}{5}$	$\frac{4}{4}$	$\frac{8}{8}$
$\frac{1}{9}$	$\frac{4}{9}$	$\frac{5}{9}$	$\frac{7}{9}$	$\frac{8}{9}$
$\frac{1}{10}$	$\frac{2}{9}$	$\frac{3}{8}$	$\frac{4}{7}$	$\frac{5}{6}$

Unit 12 Fractions

Self-check

See how much you know!

 I can do this.

 I can do this, but I need to keep trying.

 I can't do this yet.

What can I do?			
1 I can understand and explain the relationship between the whole and the parts of fractions of shapes and objects.			
2 I can understand and explain what each part of a fraction represents.			
3 I can understand that $\frac{3}{4}$ is 3 parts of 4 equal parts.			
4 I can find $\frac{3}{4}$ of a small number of objects by sharing them into 4 equal groups and counting 3 of those groups.			

I need more help with:

_____

_____

_____

Can you remember?

Fill in the missing numbers.

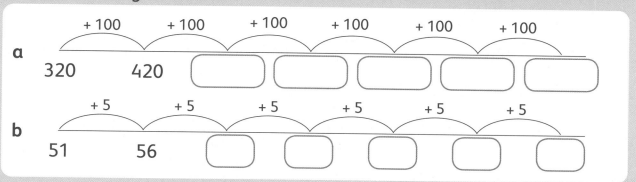

Hundreds, tens and ones

1 Complete each part-whole model.

a

987

b

897

c

789

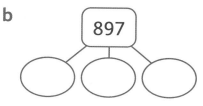
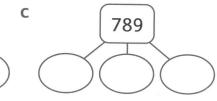

2 Show 2 ways to regroup 326.

3 Complete each part-whole model.

a

b

c

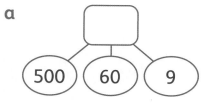
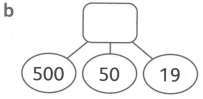
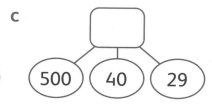

500 60 9   500 50 19   500 40 29

Comparing and ordering numbers

1 **a** Draw an arrow to show each number on the number line.

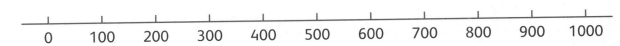

150,   790,   401,   25

0   100   200   300   400   500   600   700   800   900   1000

b Complete these statements using **<** or **>**.

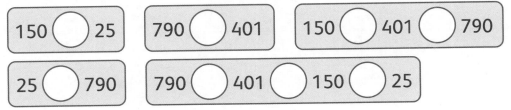

150 ◯ 25   790 ◯ 401   150 ◯ 401 ◯ 790

25 ◯ 790   790 ◯ 401 ◯ 150 ◯ 25

2 Put each set of numbers in order from least to greatest.

a 155   515   115   151   551   511

◯ < ◯ < ◯ < ◯ < ◯ < ◯

b 99   999   9   79   979   799

◯ < ◯ < ◯ < ◯ < ◯ < ◯

c 30   301   103   113   13   100

◯ < ◯ < ◯ < ◯ < ◯ < ◯

3 Look at each number below. Sort them from greatest to least.

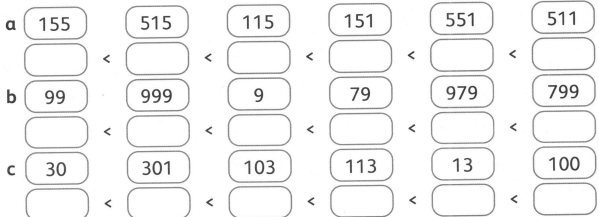

100s	10s	1s
1	2	5
3	1	4
	9	9
5	4	0
5	1	8
2	0	0

Greatest   Least

◯ > ◯ > ◯ > ◯ > ◯ > ◯

4 Use each digit card once to make the statement true.

8 ◯ 8 < ◯ 8 ◯ < 8 ◯ 8

8   9   8   9

Rounding to the nearest 10 or 100

1 Draw an arrow to show each number on the number line.
Shade which 10 it rounds to.

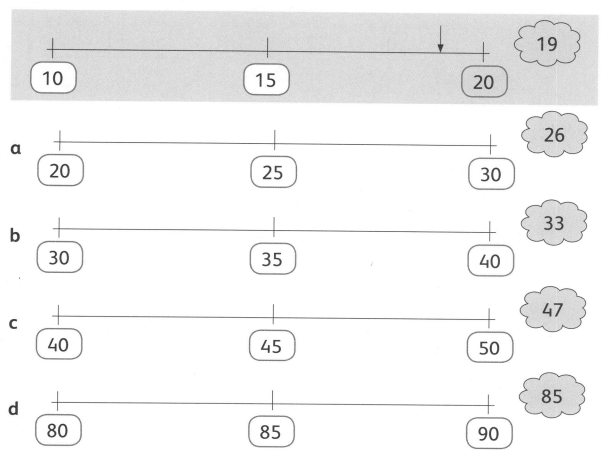

2 Round each number to the nearest 10.

a 139 ⟶ ☐   b 179 ⟶ ☐

c 299 ⟶ ☐   d 301 ⟶ ☐

e 454 ⟶ ☐   f 555 ⟶ ☐

3 Round each number to the nearest 100.

a 139 ⟶ ☐   b 179 ⟶ ☐

c 299 ⟶ ☐   d 301 ⟶ ☐

e 454 ⟶ ☐   f 555 ⟶ ☐

Patterns

1 Complete each counting pattern.

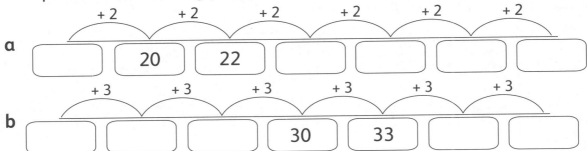

a

+2 +2 +2 +2 +2 +2

| | 20 | 22 | | | | |

b

+3 +3 +3 +3 +3 +3

| | | | 30 | 33 | | |

2 Complete the patterns and write the counting rules.

Starting number							End number	Counting rule
a	100	95	○	○	○	○	○	−5
b	101	96	91	○	○	○	○	
c	211	206	○	○	○	○	○	
d	120	○	○	○	○	○	○	−3
e	101	○	○	○	91	○	○	

3 Will 50 be shaded on the number grid?
Predict first. Then shade the numbers to check.

1	2	3	4	5	6	7	8	9	10
11	12	13	14	15	16	17	18	19	20
21	22	23	24	25	26	27	28	29	30
31	32	33	34	35	36	37	38	39	40
41	42	43	44	45	46	47	48	49	50

Predict: Will 50 be shaded? Yes ☐ No ☐

Check: Was 50 shaded? Yes ☐ No ☐

Unit 13 Patterns, place value and rounding

Self-check

 I can do this.

 I can do this, but I need to keep trying.

 I can't do this yet.

See how much you know!

What can I do?			
1 I can compose and decompose 3-digit numbers to identify each place value position of the numbers.			
2 I can decompose 3-digit numbers in different ways.			
3 I can compare and order a set of 3-digit numbers.			
4 I can round 3-digit numbers to the nearest 10 or 100.			
5 I can extend number sequences and explain the rule.			
6 I can continue a pattern and explain the rule.			

I need more help with:

_____

_____

_____

Unit 14 Addition and subtraction

Can you remember?

Convert to $ and cents.

a 150c = $□ and □c **b** 250c = $□ and □c **c** 350c = $□ and □c

151c = $□ and □c   251c = $□ and □c   450c = $□ and □c

152c = $□ and □c   252c = $□ and □c   550c = $□ and □c

Finding the value of missing numbers

1 Write the missing prices on the labels.

 **a** $2 +  = $20

What do you notice about part **b**? How does this help with the answer?

_____

_____

_____

 **b** +  = $60

2 Find the missing numbers.

a 100 − ☺ = 75   ☺ = _____

b 70 − ☁ = 10   ☁ = _____

c 20 − ⬠ − ⬠ = 10   ⬠ = _____

3 What is the value of ▱ and ★?

 ▱ = _____   ★ = _____

72

Adding multiples of 10 and 100

1 Write the missing numbers in the boxes.

a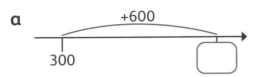

b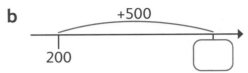

c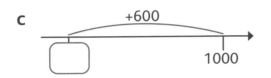

d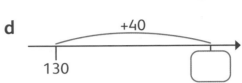

e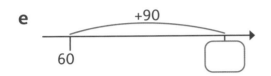

f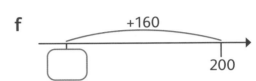

2 Complete these grids.

a

+	400	200	300
400		600	
600			
500			
	600		

b

+	120	90	110
80			
40			
50			
	190		

3 **a** Maris adds a pair of 2-digit numbers. The total is 150.
Both numbers are multiples of 10.
What could they be? Find 2 solutions.

☐ and ☐ = 150   ☐ and ☐ = 150

Can you find more solutions?
How many solutions could
there be altogether?
Write your answer in part **b**.

b _____

_____

_____

Adding pairs of 3-digit numbers

1 Complete these. Choose your own method of calculation.

	Calculation	Estimate	My working
a	253 + 135		
b	253 + 329		

2 Use all 6 digit cards each time to make two 3-digit numbers.

a Make an addition calculation that gives the largest possible 3-digit total.
Complete your calculation.

b Make another addition calculation to give the smallest possible 3-digit total.
Complete your calculation.

100s	10s	1s
⬜	⬜	⬜
⬜	⬜	⬜
⬜	⬜	⬜

+

100s	10s	1s
⬜	⬜	⬜
⬜	⬜	⬜
⬜	⬜	⬜

+

3 Solve these problems. Choose your own method of calculation.

	Problem	My working
a	262 people book for a train ride. There are 124 places left to fill. How many people can go on the train ride in total?	
b	A parcel has a mass of 235 g. A second parcel is 148 g heavier. What is the mass of the second parcel?	
c	A length of string is 345 cm. Another piece is 172 cm longer. What is the length of the longer piece of string?	

Subtraction with 3-digit numbers

1 Complete these. Choose your own method of calculation.

	Calculation	Estimate	My working
a	248 – 136		
b	253 – 128		

2 Each person is on a different journey.
Work out how far they have left to travel.

Person	Total distance of journey	How far person has travelled so far	Distance left to travel
A	275 km	152 km	
B	375 km	138 km	
C	325 km	173 km	
D	325 km	282 km	

3 Use all 6 digit cards each time to make two 3-digit numbers.

a Make up a subtraction to give the largest answer. Complete your calculation.

b Make up a subtraction to give the smallest answer. Complete your calculation.

100s	10s	1s
☐	☐	☐
☐	☐	☐
☐	☐	☐

100s	10s	1s
☐	☐	☐
☐	☐	☐
☐	☐	☐

Calculating with money

1 Write the totals in dollars and cents.
Then work out the change from $2.

	Change from $2
30c + 40c + 50c = $☐.☐☐	
31c + 41c + 51c = $☐.☐☐	
40c + 50c + 60c = $☐.☐☐	
39c + 49c + 59c = $☐.☐☐	

2 Look at the ticket prices for the local cinema and theatre.

Event	Adult	Child
	$8	$5

Event	Adult	Child
THEATRE	$11	$4

Colour in the cheaper option each time.

a (3 adults to the cinema) or (2 adults to the theatre)

b (2 adults to the theatre) or (3 children to the cinema)

c (5 adults to the cinema) or (4 adults to the theatre)

d (3 adults and 2 children to the cinema) or (2 adults and 2 children to the theatre)

3 Draw a line to join the calculation with the best estimate.

a	$1.99 + $2.99
b	$1.49 + $1
c	$5 + $2.49
d	$5.99 + $9.99
e	$2.49 + $2.49 + $4.99

$2.50
$5
$10
$7.50
$16

4 Look at the hot drinks prices.

Tea 75c   Coffee 90c   Hot chocolate $1.10

a How much change from $5, when a customer buys:

1 tea and 1 coffee?   Change: _____

2 teas and 1 coffee?   Change: _____

1 tea and 2 coffees?   Change: _____

2 teas and 2 coffees?   Change: _____

1 of each drink?   Change: _____

b How much more money is needed to buy 2 of each drink?

_____ cents

77

Unit 14 Addition and subtraction

Self-check

See how much you know!

 I can do this.

 I can do this, but I need to keep trying.

 I can't do this yet.

What can I do?			
1 I can work out the value of an unknown quantity in an addition calculation and in a subtraction calculation.			
2 I can identify complements of 100 from sets of numbers.			
3 I can add pairs of multiples of 10 and 100.			
4 I can estimate to check the answers to addition calculations and subtraction calculations.			
5 I can add pairs of 3-digit numbers.			
6 I can take away a 3-digit number from a 3-digit number.			
7 I can use notes and coins to pay for items and work out the change.			

I need more help with:

_____

_____

_____

Can you remember?

Estimate the length of each line. Then measure the lines using a ruler.

a Estimate [ ]

Measurement [ ]

b _____

Estimate [ ] Measurement [ ]

c _____

Estimate [ ] Measurement [ ]

d ____

Estimate [ ] Measurement [ ]

Time

1 Write the missing time on each number line.

a   + 10 min + 10 min + 10 min

3:15 [ ]

b   + 10 min + 10 min + 10 min

[ ] 9:55

c   + 10 min + 10 min + 10 min

4:25 [ ]

d   + 10 min + 10 min + 10 min

[ ] 9:10

e   + 10 min + 10 min + 10 min

4:20 [ ]

2 How long from start to finish for these TV shows?

Start	Finish	Length
10:30 a.m.	11:00 a.m.	30 minutes
10:30 a.m.	11:05 a.m.	
9:35 a.m.	10:00 a.m.	
9:35 a.m.	10:10 a.m.	
9:45 a.m.	10:10 a.m.	
10:10 p.m.	11:05 a.m.	
7:55 p.m.	8:50 p.m.	
7:45 p.m.	8:50 p.m.	

3 Steps to bake a loaf of bread:

- Wash hands and work area – 5 minutes.
- Use recipe and measure ingredients – 5 minutes.
- Mix flour, yeast, sugar, salt and water – 5 minutes.
- Knead dough – 10 minutes.
- Cover and leave to rise – 30 minutes.
- Bake – 30 minutes.
- Leave to cool – 10 minutes.

If you need fresh bread at 12:30, what time will you start making it?

_____

_____

_____

_____

Perimeter of 2D shapes

1 Find the perimeter of each shape.

a

3 cm   3 cm

3 cm

$3 + 3 + 3 = \boxed{}$ cm

b

5 cm

5 cm   5 cm

5 cm

$\boxed{} + \boxed{} + \boxed{} + \boxed{} = \boxed{}$ cm

c

6 cm

1 cm   1 cm

6 cm

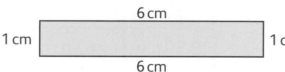

$\boxed{}$ cm

2 Annay wants to put netting around his vegetable patch. How many metres of netting will he need?

$\boxed{}$ m

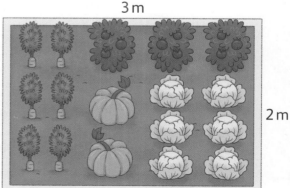

3 m

2 m   2 m

3 m

3 Draw 3 different shapes with a perimeter of 8 cm each.

Area

1 What is the area of each shape?

a [ ] squares   b [ ] squares

c [ ] squares   d [ ] squares

2 A builder is tiling a kitchen floor. Each tile is 1 square unit. How many tiles will he need?

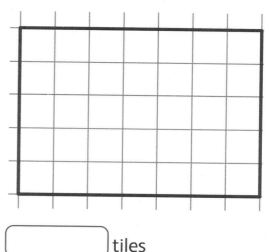

[ ] tiles

3 Draw 3 different shapes with an area of 6 square units each.

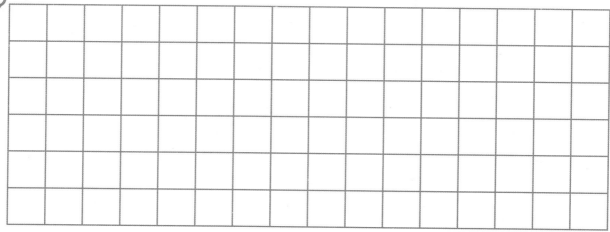

Unit 15 Time and measurement

Self-check

See how much you know!

 I can do this.

 I can do this, but I need to keep trying.

 I can't do this yet.

What can I do?			
1 I can work out the interval (amount of time) between 2 given times.			
2 I can measure the perimeter of 2D shapes.			
3 I can measure the area of a grid in square units.			
4 I can draw rectangles and measure the length of each side to find the perimeter.			

I need more help with:

_____

_____

_____

Can you remember?

Write 2 multiplication and 2 division sentences to match each array.

a

[ ] × [ ] = [ ]

[ ] × [ ] = [ ]

[ ] ÷ [ ] = [ ]

[ ] ÷ [ ] = [ ]

b

[ ] × [ ] = [ ]

[ ] × [ ] = [ ]

[ ] ÷ [ ] = [ ]

[ ] ÷ [ ] = [ ]

Multiplication and division facts

1 Complete these multiplication grids.

a

×	3	6	9
4	12		
5			
8			

b

×	7	6	9
2			
	7		
10			

c

×		3	
		18	
9	81		
8			40

2 Write the missing information.

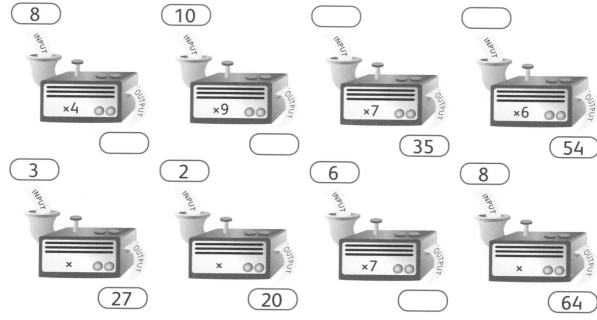

8 → ×4 → [ ]

10 → ×9 → [ ]

[ ] → ×7 → 35

[ ] → ×6 → 54

3 → × → 27

2 → × → 20

6 → ×7 → [ ]

8 → × → 64

3 Use a set of 1 to 8 digit cards to make four 2-digit numbers.
The challenge is to have one 2-digit number in each multiplication table.

3× table

4× table

6× table

8× table

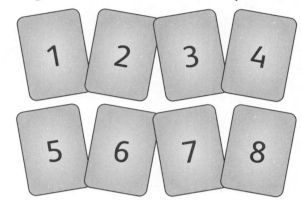

Multiplying 2-digit numbers

1 Complete these.

	Calculation	Estimate	Use the written column method
a	32 × 3		
b	33 × 5		
c	28 × 4		

2 Find the value of each circle by multiplying the numbers in the 2 circles below.

a 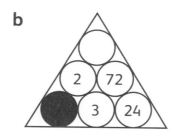 b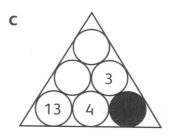

c

3 Find the missing digits.

a

100s	10s	1s
	2	☐
×		3
	7	2

b

100s	10s	1s
	2	☐
×		4
☐	1	2

c

100s	10s	1s
	7	1
×		☐
☐	5	5

4 Jack and Zara have done some multiplications.
a Use estimates and other facts you know to decide if their answers could be correct. Mark it with a ✓ or a ✗.

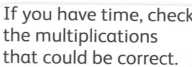
If you have time, check the multiplications that could be correct.

	✓ or ✗
19 × 4 = 86	
25 × 5 = 125	
53 × 2 = 96	
42 × 3 = 126	
28 × 4 = 122	
23 × 3 = 59	

b Choose a multiplication that you marked wrong (✗).
Explain why you know it is wrong.

_____

_____

_____

Solve this problem.
A shop sells 4 boxes of 32 red pencils.
It also sells 3 boxes of 43 blue pencils.

a Did the shop sell more blue pencils than red pencils?   **Yes   No**

b How many more pencils of one colour did it sell?

Division with and without remainders

  Complete these.

Calculation	Estimate	Use a mental method or a written method
a 63 ÷ 3		
b 76 ÷ 4		
c 85 ÷ 5		

2 Complete these divisions.

a 19 ÷ 3 =

b 42 ÷ 4 =

c 43 ÷ 5 =

d 17 ÷ 2 =

e 28 ÷ 3 =

f 39 ÷ 4 =

3 Match each division to a remainder.

(12 ÷ 5) (21 ÷ 5) (13 ÷ 5) (36 ÷ 5) (27 ÷ 5) (48 ÷ 5)

(1) (2) (3)

4 Solve this problem.
Viti has fewer than 20 toys. She puts them in groups of 3.
There is 1 toy left. She then puts them in groups of 4.
There are no toys left. How many toys does Viti have?

[ ] toys

My working

5 David and Maris have done some divisions.
a Use estimates and other facts you know to decide if their answers
could be correct. Mark it with a ✓ or a ✗.

> If you have time, check the multiplications that could be correct.

	✓ or ✗
38 ÷ 2 = 14	
24 ÷ 3 = 7 r2	
76 ÷ 4 = 19	
49 ÷ 2 = 25	
86 ÷ 5 = 17 r1	
61 ÷ 3 = 20 r2	

b Choose a division that you marked wrong (✗).
Explain why you know it is wrong.

_____

_____

Unit 16   Multiplication and division

Self-check

 I can do this.

 I can do this, but I need to keep trying.

 I can't do this yet.

See how much you know!

What can I do?			
1 I know the 1, 2, 3, 4, 5, 6, 8, 9 and 10 times tables.			
2 I know division facts that are related to known multiplication facts.			
3 I can estimate to check the answers to multiplication calculations and division calculations.			
4 I can multiply any 2-digit number by 2, 3, 4 and 5.			
5 I can divide 2-digit numbers by 2, 3, 4 and 5.			

I need more help with:

_____

_____

_____

Unit 17   Fractions

a The baker has 12 eggs.
He uses half the eggs.
How many are left?

b The farmer has 16 goats.
He sells $\frac{1}{4}$ of the goats.
How many are left?

Adding and subtracting fractions

1 Use the diagrams to help you complete each addition.

a $\frac{2}{5} + \frac{1}{5} = \frac{\square}{\square}$

b $\frac{1}{6} + \frac{4}{6} = \frac{\square}{\square}$

c $\frac{2}{5} + \frac{2}{5} = \frac{\square}{\square}$

d $\frac{5}{7} + \frac{1}{7} = \frac{\square}{\square}$

2 Complete the calculations.

a $\frac{3}{5} - \frac{1}{5} = \frac{\square}{\square}$ **b** $\frac{2}{3} - \frac{1}{3} = \frac{\square}{\square}$ **c** $\frac{4}{7} - \frac{2}{7} = \frac{\square}{\square}$ **d** $\frac{5}{6} - \frac{4}{6} = \frac{\square}{\square}$

3 Use a bar model to help you to work out these calculations.

a $\frac{\square}{5} + \frac{1}{5} = \frac{3}{5}$

$\frac{\square}{5} - \frac{1}{5} = \frac{3}{5}$

$\frac{\square}{5} + \frac{2}{5} = \frac{3}{5}$

$\frac{\square}{5} - \frac{2}{5} = \frac{3}{5}$

b $\frac{\square}{6} + \frac{1}{6} = \frac{5}{6}$

$\frac{\square}{6} - \frac{1}{6} = \frac{4}{6}$

$\frac{\square}{6} + \frac{1}{6} = \frac{3}{6}$

$\frac{\square}{6} - \frac{1}{6} = \frac{2}{6}$

c $\frac{2}{9} + \frac{\square}{9} + \frac{2}{9} = \frac{8}{9}$

$\frac{2}{8} + \frac{\square}{8} + \frac{2}{8} = \frac{8}{8}$

$\frac{2}{9} + \frac{\square}{9} - \frac{2}{9} = \frac{1}{9}$

$\frac{2}{8} + \frac{\square}{8} - \frac{2}{8} = \frac{1}{8}$

89

4 Shade each correct calculation.

$\frac{3}{6} - \frac{1}{6} = \frac{2}{6}$	$\frac{3}{6} - \frac{1}{6} = \frac{4}{6}$	$\frac{3}{6} + \frac{1}{6} = \frac{2}{6}$	$\frac{3}{6} - \frac{1}{6} = \frac{1}{6}$	$\frac{1}{6} + \frac{1}{6} = \frac{2}{6}$
$\frac{6}{7} - \frac{1}{7} = \frac{7}{7}$	$\frac{6}{7} + \frac{1}{7} = \frac{7}{7}$	$\frac{5}{7} - \frac{2}{7} = \frac{7}{7}$	$\frac{5}{7} + \frac{2}{7} = \frac{7}{7}$	$\frac{1}{7} - \frac{1}{7} = \frac{7}{7}$
$\frac{3}{4} - \frac{1}{4} = \frac{1}{4}$	$\frac{3}{5} - \frac{1}{5} = \frac{1}{5}$	$\frac{6}{6} - \frac{5}{6} = \frac{1}{6}$	$\frac{5}{7} - \frac{1}{7} = \frac{1}{7}$	$\frac{2}{8} - \frac{1}{8} = \frac{1}{8}$
$\frac{3}{5} - \frac{1}{5} = \frac{4}{5}$	$\frac{3}{5} - \frac{1}{5} = \frac{2}{5}$	$\frac{3}{5} + \frac{1}{5} = \frac{2}{5}$	$\frac{3}{5} + \frac{1}{5} = \frac{4}{5}$	$\frac{3}{5} - \frac{1}{5} = \frac{1}{5}$
$\frac{3}{7} - \frac{1}{7} = \frac{2}{7}$	$\frac{3}{7} - \frac{1}{7} = \frac{4}{7}$	$\frac{3}{7} + \frac{1}{7} = \frac{2}{7}$	$\frac{3}{7} - \frac{1}{7} = \frac{1}{7}$	$\frac{3}{7} + \frac{1}{7} = \frac{2}{7}$

5 Use each fraction once so that each row and each column has the same total.

$\frac{1}{10}$ $\frac{3}{10}$ $\frac{5}{10}$

$\frac{1}{10}$ $\frac{3}{10}$ $\frac{5}{10}$

$\frac{1}{10}$ $\frac{3}{10}$ $\frac{5}{10}$

Equivalent fractions

1 Use the diagram to list any equivalent fractions you can find.

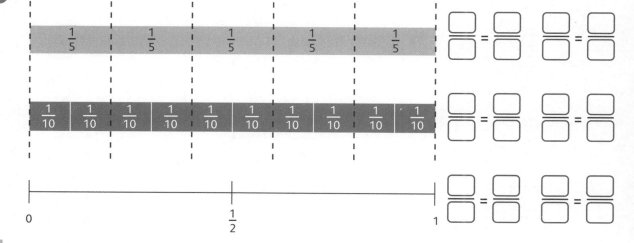

2 Join the equivalent fractions.

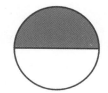 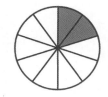   

$\dfrac{10}{10}$ $\dfrac{5}{10}$ $\dfrac{1}{5}$ $\dfrac{6}{10}$ $\dfrac{4}{5}$

3 Complete these calculations.
To help you, think about fractions that are equivalent.

a $\dfrac{\square}{4} + \dfrac{\square}{4} = \dfrac{1}{2}$ b $\dfrac{\square}{4} + \dfrac{\square}{4} = \dfrac{5}{10}$ c $\dfrac{\square}{10} + \dfrac{\square}{10} = \dfrac{2}{4}$

d $\dfrac{\square}{10} + \dfrac{\square}{10} + \dfrac{\square}{10} = \dfrac{1}{2}$ e $\dfrac{\square}{10} + \dfrac{\square}{10} + \dfrac{\square}{10} = \dfrac{4}{5}$

Comparing and ordering fractions

1 Use the symbol **<** or **>** to complete each statement.

a 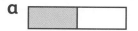    b

c    d

2 Use each card once to complete these statements.

a $\dfrac{1}{\square} > \square$ b $\dfrac{\square}{9} = \square$

c $\dfrac{1}{\square} > \square$ d $\square = \dfrac{\square}{10}$

 1   2   3   4

Unit 17 Fractions

Self-check

 I can do this.

 I can do this, but I need to keep trying.

 I can't do this yet.

See how much you know!

What can I do?			
1 I can understand and explain a fraction as being the numerator divided by the denominator.			
2 I can use a diagram to show equivalent fractions.			
3 I can add and subtract fractions with the same denominator and model them with a diagram.			
4 I can estimate the answer when adding and subtracting fractions with the same denominator.			
5 I can put a set of unit fractions in order.			
6 I can compare and order fractions with the same denominator and different numerators.			
7 I can use the symbols < or > to compare and order values.			

I need more help with:

_____

_____

_____

Can you remember?

Zara spins a 1 to 6 spinner.

Use these words:

| It will happen | It might happen | It will not happen |

How likely is it that Zara spins a 3?

How likely is it that Zara spins a number?

How likely is it that Zara spins a 14?

Carroll diagrams and Venn diagrams

1 Write 2 different numbers to go in each section.

	Multiple of 5	Not a multiple of 5
Multiple of 3		
Not a multiple of 3		

2 Write 5 different numbers in each section on the Venn diagram.

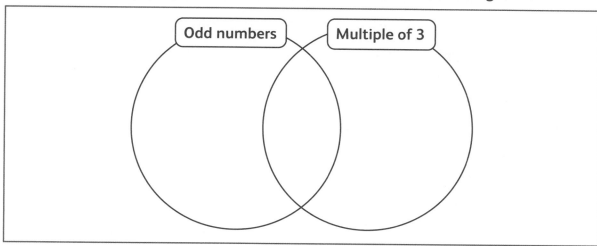

Odd numbers　　Multiple of 3

3 Play this game with a partner. Each player must choose a section on the Carroll diagram (A, B, C or D).
Spin a spinner 3 times. Then add the 3 numbers together.
Write the answer in the correct place in the Carroll diagram.
After adding 12 numbers, see which player has the most numbers in their section.

	10 or less	Not 10 or less
Even	A	B
Not even	C 9	D

Lists, tables, charts and graphs

1 Jack looks for minibeasts in his garden every weekend.
He has recorded some of his results in the table.

The number of minibeasts Jack saw each week

Week	Worms	Snails	Total
Week 1	IIII	II	6
Week 2		I	11
Week 3	IIII	IIII II	
Week 4	III		9
Week 5		III	8

a Complete the table.

b In which week did Jack see the most minibeasts?

c In which week did Jack see the least minibeasts?

d How many more minibeasts did Jack see in Week 2 than in Week 1?

e Complete the bar chart to show the total number of minibeasts Jack saw each week.

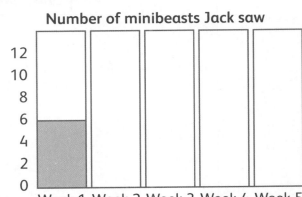

Number of minibeasts Jack saw

12
10
8
6
4
2
0
Week 1 Week 2 Week 3 Week 4 Week 5

Chance

1 Jack and Annay are playing a fishing game.
There are 12 magnetic sea creatures in a box:

- 4 blue fish
- 4 crabs
- 4 seahorses

They take turns to pick out a creature. Then they put the creature back.

Use these words to answer the questions:

> **It will happen**

> **It might happen**

> **It will not happen**

a How likely is it that Jack will pick out a fish?

b How likely is it that Jack will pick out a sea creature?

c How likely is it that Jack will pick out a robot?

d How likely is it that Jack will pick out a crab?

Can you add one **chance** question of your own? Write it in part **f**.

e How likely is it that Jack will pick out a strawberry?

f _____

Unit 18 Statistical methods and chance

Self-check

See how much you know!

 I can do this.

 I can do this, but I need to keep trying.

 I can't do this yet.

What can I do?			
1 I can organise information into a list, a table and a chart.			
2 I can record data using a tally chart.			
3 I can answer questions about a pictogram and a bar chart.			
4 I can sort objects and shapes in a Carroll diagram.			
5 I can answer questions about Venn diagrams and Carroll diagrams.			
6 I can interpret and explain the data presented in tables, bar charts and pictograms.			
7 I can describe the chance of an event happening.			
8 I can describe the results of a chance experiment.			

I need more help with:

_____

_____

_____